PROF CHRIS BAILEY

# A Designer's Guide to Simulation with Finite Element Analysis

Author:

**Vince Adam**

ISBN 978-1-874376-3





**ACKNOWLEDGEMENT**

I wish to take this opportunity as Chairman of NAFEMS Education and Training Working Group to thank the Working Group members for their help and support in the preparation of this book. The composition of the Working Group is:

| | |
|---|---|
| Adib Becker | University of Nottingham |
| David Ellis | IDAC |
| Peter Gosling | University of Newcastle |
| Trevor Hellen | Consultant |
| Herve Morvan | University of Nottingham |
| Derek Pashley | Consultant |
| Anup Puri | SELEX sas |
| Bryan Spooner | Consultant |
| Jim Wood | University of Strathclyde |
| Rodney Dreisbach | The Boeing Company |
| Yasar Degar | Hochschule Rapperswil |
| Jack Reijmers | Nevesbu |
| Stefan Kukula | Crowcon Detection Instruments |
| Grant Steven | STRAND7 Software |
| Jesus Lopez Diez | University Polytechnic Madrid |
| John Smart | Former ETWG chairman |

Prof. Adib Becker
Chairman
NAFEMS Education and Training Working Group

**Disclaimer**

*Whilst this publication has been carefully written and subject to peer group review, it is the reader's responsibility to take all necessary steps to ensure that the assumptions and results from any finite element analysis which is made as a result of reading this document are correct. Neither NAFEMS nor the authors can accept any liability for incorrect analysis.*

## AUTHOR'S ACKNOWLEDGEMENT

I would like to again thank NAFEMS for the opportunity to compile this guideline for design analysts. Equally important, I would like to thank my colleagues at SolidWorks and other friends "in analysis" for their input on this document as well as our customers who have put their trust in me and have allowed me to grow as an educator in this field. While having ample opportunity to learn from my own mistakes, helping others learn from theirs has been one of the more rewarding aspects of my career. Thus it is only with great respect that I have shared some of those lessons in this book to help readers avoid re-inventing the wheel. Finally, I would like to thank my family for letting me take precious time away from them to work on this so shortly after completing a previous NAFEMS project.

# Contents

# 1. Introduction

This book is an overview of methods and best-practices for mechanical design engineers and designers who are using or plan to utilize Finite Element Analysis (FEA) to validate design concepts or predict and correct product failure. The target readership is part-time users of CAD (Computer Aided Design) embedded analysis packages or those whose interface relies primarily on geometric references. This should not imply that the information presented herein is not important to users of other, more complete, analysis packages or that these users are by default any more capable of successfully simulating product performance. The assumption, right or wrong, is that with the greater investment required by purchasers of more complete tools, comes more investment in training and time to learn proper modelling practices. On the other hand, many users of design analysis tools receive little training beyond a basic software introduction, if even that. These design engineers or designers are typically reliant on whatever educational material is provided by their vendor and much of this is tool related. This book should pick up where that material left off.

## 1.1 Should Designers Be Doing FEA?

Articles in the media and marketing from the software vendors suggest that anyone capable of learning CAD can and should perform FE based simulation on their products. However, in some circles, primarily consisting of more full-time analysts, the concept of FEA for the masses is met with scorn and some go so far as to suggest that putting simulation in the hands of non-specialists is akin to handing a child a loaded gun. As with many things in life, reality is somewhere in the middle of these extreme viewpoints.

Addressing the first point, it is important not to lose sight of the fact that FEA is still an engineering tool, requiring engineering level insight into failure and material characteristics and engineering decisions based on the results. The results of even a properly constructed simulation are not always what they seem. As will be discussed later in this book, data reported by a finite element solution requires interpretation and this is a skill that goes beyond one's ability to use the software. A user that denies or shirks responsibility for understanding the product, material, and/or software specific variability in any given scenario probably shouldn't be using analysis. This is why podiatrists don't typically perform open-heart surgery. Suffice it to say that any product development professional that is willing to learn the technology, in addition to the tool, can obtain value from simulation. Simply learning the software isn't enough.

Is leaving FEA to the specialists the answer? There are two important reasons why the product development industry will never revert to that paradigm. First of all, in a quest to drive innovation into products and reduce time to market, initial decisions in the design process need to be validated at the earliest possible opportunity and lessons learned need to be incorporated immediately. The obvious derivative from this line of thinking is that the people making those decisions need to be enabled with tools to enable validation. Currently FE based simulation is the tool best positioned to provide this. Secondly, the swing towards designer or design engineer simulation has started and has so much momentum, it may be irreversible even if it didn't have merit (and it does) in light of the big picture. Design engineers are already convinced they can and should be doing FEA and many feel, rightly or wrongly, that they are fairly successful with it. It would not be a trivial task to convince even the ones using it poorly that they should give it up.

Consequently, the question, "Should designers be doing FEA?" is moot. They are doing FEA and their numbers will continue to swell as additional physics such as fluid flow simulation, electro-magnetics (E/M), and manufacturing simulation tools are incorporated into the various CAD packages and the software continues to become easier to use. The price of entry, historically a barrier to widespread FEA use in the design ranks, is also dropping dramatically, even disappearing, as some CAD packages incorporate limited analysis capabilities at no charge. Many major software providers are committed to this direction and it is inconceivable that this will change.

There is an important caveat to the above conclusion. While the analysis performed by a part-time user is important, even critical, to the improvement of innovation and speed in the design process, it shouldn't replace traditional validation. When the cost or risk of proceeding to parts or tooling is great, it is recommended that a more experienced analyst be consulted to either review the design analysis conclusions or repeat the analysis with more full-featured tools, if that adds value. A comprehensive project report will help facilitate this "final check."

## 1.2 Purpose of This Book

Consistent with their founding mission, "To promote the safe and reliable use of finite element and related technology," NAFEMS welcomes the growing ranks of "Design Analysts." To aid them in making every simulation meaningful, this text is targeted at filling the gaps in knowledge left by basic software training. This book will attempt to summarize the collective body of knowledge as it pertains to proper modelling methods, analysis assumption construction, and results interpretation. Realistic expectations for the value of design analysis will be set that, while possibly tempering the blind enthusiasm some get when they are first exposed to FEA, should not discourage anyone who is serious about getting it right versus getting it done.

This is not a theoretical exploration of finite elements or their mathematical basis but a practical guide to using the technology effectively. The primary focus will be on CAD-embedded analysis tools or those that exclusively rely on geometric references for meshing and boundary condition input. However, much of the material will also be relevant to design engineers using more full-featured FEA systems. Where applicable, the limitations of the current offering of CAD-driven tools will be noted so that purchasers of more complete systems can justify their decision or investigate augmenting their more cumbersome tools with analysis products that integrate better with their CAD. Similarly, users of more limited systems can determine if they should be considering an upgrade once they understand the full potential of the technology. Based on the development of the technology over the last 5 years, it is reasonable to speculate that much of this discussion will still be applicable 5 years out. However, should a new paradigm or breakthrough technology, this author will be more than happy to offer a retraction!

# 2. The Role of Simulation in Product Development

FEA software comes in all price ranges, from free to almost prohibitive. It also comes in many forms ranging from embedded in and exclusive to a CAD system to open tools allowing nearly limitless interchange of geometry and mesh data between various CAD systems and FE interfaces and solvers. A common misconception is that the lower the cost and the more CAD-friendly a tool is, the lower the impact it can have on the design process, warranting a less thoughtful attitude for model setup and results interpretation. However, in most cases, the solver technology and the basic model building capabilities in the more "friendly" tools are the same or similar to the higher end tools. What makes a basic tool 'basic' is that the user is limited to a smaller subset of capabilities available in more complete tools. However, if a product's performance can be simulated within that limited subset of tools, the results and subsequent impact on design decisions is equally meaningful. Therefore, the success of a CAD-embedded analysis tool is controlled by three related factors:

1. Modeling geometries and behaviors within the limitations of the tool
2. Proper or sufficient implementation of the tool with respect to timing, training, and expectations
3. A sufficient understanding of the relevant failure modes and results quantities, both in the software and in the products being simulated

The first factor will be addressed indirectly in subsequent chapters as the limitations of CAD-embedded tools are put into their proper perspective. Addressing the second and third factors will comprise the majority of this text, in a general sense within this chapter and then more specifically afterwards.

## 2.1 Applicability of Simulation At Various Stages In Design

Simulation is typically integrated into the design process at one of three stages:

1. Failure Verification
2. Design Verification
3. Concept Verification

Most companies begin by utilizing FEA to address known problems or failures in the *Failure Verification* mode. In short, something broke that they weren't expecting to break. The software is used to provide insight into why it broke and, hopefully, how to fix it. After a certain level of success, companies come to the realization that if they could show why a part or system failed analytically after it failed, they should be able to predict if a design is going to fail beforehand. At this stage, the design process hasn't been altered significantly. An additional task has been inserted in between the Design and the Prototype stage. In these situations, the design has progressed using the same tools, techniques and insights as it would have prior to the availability of simulation. This completed design is simply validated in the virtual world, and corrected if necessary, before parts are cut. This would be called *Design Verification.*

If a company stopped at this point and simply continued to use simulation as a Design Checker, there is still value in the effort. However, some companies have taken that final step in the integration of simulation technology by inserting it into the decision making process that leads to a completed design. This allows them to validate any decision, be it material choice, number of fasteners, quantity and placement of ribs, nominal wall, etc., before committing to subsequent decisions which may depend on the first or at least interact with it. Thus ideas are verified in their conceptual stage, hence the term, *Concept Verification.*

Consider this scenario... In plastic part design, the nominal wall thickness drives the size of many features. Rib height, width and corner radii are all tied through ratios to the nominal wall. Consequently, with an inappropriate or sub-optimal choice for nominal wall thickness, typically one of the first decisions in the CAD layout of a plastic component, many other decisions are committed. Should an FEA based validation, or even worse, a tooled prototype of the completed design, show that the nominal wall was under- or over-specified, the amount of rework might be significant. If the problem was caught in a Design Verification analysis, it is comforting for one to think that a mistake has been avoided through the diligence of FEA. However, from a product schedule basis, days to weeks of CAD modelling time may be required to update all the features driven by this nominal wall change, not to mention the changes to product cost if the wall needed to be increased. Cost over-runs can then lead to re-evaluation of fastener or other component decisions and so on until the program grinds to a halt and gets shelved.

## 2.2 Business Benefits of Early Simulation

While the previous scenario may be extreme, it is not unthinkable. In most cases, the cost to a company for making sub-optimal decisions early in the design process is much less dramatic or visible. From a more practical standpoint, a company and, consequently, its design engineers receive benefit from FEA when it impacts one of the following objectives in any design program:

- Reduced Product Development Time;
- Increased Innovation;
- Reduced Product Cost;
- Reduced Development Cost;
- Improved Product Quality.

In essence, one can say that using FEA mitigates the *risk* associated with relying on physical prototypes, prior experience, or luck to address these objectives. The weight each of these objectives are given as they relate to total project profitability varies from company-to-company and product-to-product but each are important and must be controlled and optimized to a certain extent.

### 2.2.1 Reduced Product Development Time

Time-to-Market has been identified by product development experts as mission critical for companies in competitive industries. The importance of this isn't always shared with design engineers despite the fact that they are key players in the setting and keeping of project schedules.

Simulation used as a Failure Verification tool can help speed recovery when a problem is found at the prototype or pre-production stage. While there is certainly value in this, most managers perceive that they would have been better off if they hadn't needed to use analysis in the first place since its use added even more time to the release date overrun. In these scenarios, the value analysis brings to the table isn't always recognized.

When used as a Design Verification tool, simulation starts to get some credit for providing value, mostly in problem avoidance. Again, the fact that it is pointing out mistakes, albeit while helping to avoid others, cannot be ignored. If problems are not indicated by the design verification simulation, the value of the analysis is typically perceived as null to positive if the schedule was not impacted by the task or as a negative if the schedule was.

When used as a Concept Verification tool, FEA has the opportunity to make the biggest impact on development time for a number of reasons. Analyses performed on conceptual models tend to set up and run more quickly as a result of less complex CAD models. More importantly decisions can be validated before cascading dependencies are implemented, thus reducing rework time later on in the project...the cost of which is typically very high.

### 2.2.2 Increased Innovation

A company's ability to innovate seems to be edging out time-to-market as the most celebrated design objective in product development literature. With ever-increasing production efficiencies and techniques, combined with the challenge of global competition from companies who may be able to produce a known technology more quickly for less money, coming to the market with new technology may often mean the difference between survival and failure in today's economy.

Only using simulation in the Concept Verification mode has much of an impact on the innovativeness of a company. If designs progress the way they always had and simply verified, as with a spell checker, the resulting product will likely be similar to the ones that preceded it. However, if FEA is used as a conceptual tool to investigate new materials, manufacturing methods, or component interactions that challenge accepted design practices, innovation will be a natural by-product and optimization becomes much more practical.

### 2.2.3 Reduced Product Cost

Using a similar argument as above, most of a product's cost is committed in the earliest stages of concept development. Once a design has made it to the stage where Failure Verification is required, cost may go in but it is rarely coming out. If Design Verification happens with some time left in the schedule, limited optimization might happen, thus allowing FEA to contribute to cost reduction. Using FEA to optimize cost by optimizing decisions early in the design process, again in the Concept Verification mode, provides the best opportunity for the technology to impact final product cost.

### 2.2.4 Reduced Development Cost

Reduction in product testing is often cited as one of the top goals for implementing FEA in the design process. Most engineering groups can tally up the cost of prototype iterations, including part cost and testing expenses. One automotive radiator manufacturer estimated that wind tunnel tests for a new product, across all the vehicles that it might be installed in, would exceed $2 Million. That same manufacturer was able save the majority of that cost by utilizing CFD in a Design Verification mode to determine which configuration represented the extreme case and they were able to justify a single test to their client based on this. It is unrealistic to think that all product testing may be eliminated through simulation. However, reducing the number of iterations has a major impact on product development cost. Additionally, simulation in the Concept Verification can eliminate alternatives from the decision matrix that might have otherwise required testing to explore. More importantly, FEA can explore alternatives, or software prototypes, that time or funds would never be made available to explore. This equates to more opportunities for innovation.

### 2.2.5 Improved Product Quality

Many factors go into the quality aspect of a product. Some can be addressed and improved with simulation. Understanding what quality means to your company as well as your customer base can help focus design decisions and efforts. One power tool company prided themselves on their durability and dependability. Their design efforts tended to focus on motor and brush life, gear train life, and other traditional wear components. Simulation didn't play a large role in their development process since it couldn't address these issues. Recently, they've come to realize that these traits, which were so important to tradesmen 20 years ago, aren't as marketable in our more disposable oriented economy. Their customers are now more focused on maximizing performance and surviving single-use abuse scenarios, such as drops from a ladder. The users will most likely shop for the latest technology long before wear becomes an issue. All of a sudden, FEA has become important to them since it can address housing needs of higher torques and drop events. They are scrambling to choose tools, users, and an implementation strategy.

Using FEA in a Concept Validation mode can have the biggest impact on product quality. Again, the decisions that most affect the final product performance and quality are typically made in the earliest stage of design. While Design and Failure Verification may address inherent quality deficiencies, they don't necessarily improve quality on their own.

## 2.3 Capabilities and Limitations of FEA

One point made in the previous discussion about quality was that the referenced power tool manufacturer acknowledged that FEA wasn't necessarily going to address motor brush life or gear noise. The time has not yet come where a design engineer can take a complete CAD assembly model and specify operational inputs such as battery voltage or engine horsepower and expect to see a system behave virtually as it would in physical testing. The current state of the technology still requires engineers to structure a problem within the limitations of their *problem-solving environment.*

### 2.3.1 The Problem-Solving Environment

A designer or design engineer's personal problem-solving environment is a fluid set of resources that enables him or her to make qualified decisions on engineering issues that arise. As an engineer's experience base broadens, so does their problem-solving environment. Their personal problem-solving environment includes their:

- Knowledge of the inputs (material, geometry, loads, history);
- Knowledge of the allowables (stress, displacement, frequencies);
- Knowledge of the physics involved (plasticity, fatigue, dynamics);
- Knowledge of past performance of similar systems;
- Tools available to gather data (CAD, FEA, Testing);
- Knowledge and limitations of said tools;
- Availability of resources to supplement the above.

When a design engineer attempts a solution that exceeds the bounds of their problem-solving environment, risk begins to accumulate. Remembering that the value of FEA in the design process is to minimize risk, knowingly exceeding one's problem solving environment somewhat negates the point of using FEA in the first place. It is more commonplace that an engineer isn't aware of their personal limitations or, oftentimes, misjudges their limitations. It is reasonable to conclude that a design engineer working within their problem-solving environment is more effective than one who tries to exceed their environment. Diligent peer review, both of specific projects and general processes can aid in minimizing the risk when the bounds of a team's problem-solving environment are pushed.

A more detailed discussion of all aspects of the problem-solving environment is beyond the scope of this book. However, one should keep this concept in mind as the more FEA-specific aspects of problem solving are discussed. Suffice it to say that when a design engineer chooses to use FEA as a tool for improved decision-making, whether it be in a Failure, Design, or Concept Verification mode, he or she should be cognizant of their limitations.

### 2.3.2 Garbage In-Garbage Out

Specific to FEA, remember that a finite element solver simply processes data to provide a solution. It doesn't possess any additional insight, can't prompt a user to consider other options, or in most cases, point out flawed inputs. In a manner of speaking, your FEA solver trusts that you know what you're doing. To many newer users, this may be one of its biggest limitations.

### 2.3.3 Precise Answers to Imprecise Questions

As will be shown in the later chapters of this book, a user rarely has the opportunity to input 'accurate' loads, properties, and geometry. In nearly all cases, these inputs will be a user's best guess based on nominals, worst-case scenarios, or average operating conditions. However, the FEA solver deals in absolutes, not approximate values. Just as in linear math where 2 + 2 equals exactly 4, a given force on a given geometry with a given material stiffness will yield a very exact solution. This solution will be reported with several digits after the decimal point and for all intents and purposes will appear authoritative and exact. Unfortunately, when the inputs are best-guess assumptions, this 'exact' looking response can easily mislead a user into an erroneous design decision.

### 2.3.4 Discretization Error

The very nature of FEA is to break a continuum into small sub-parts called elements to make the math reasonable so a solution can be reached efficiently. The validity of the solution is dependent upon the size, shape, and type of the elements used. If a user isn't cognisant of the error that can get introduced with improper element usage, then he or she may not realize that reasonable looking results may be far off the mark. Tools exist in most design analysis codes to aid in checking mesh validity. If these aren't available, a manual investigation of appropriate mesh sizing will still be required. This is an important yet often overlooked step.

## 2.4 Defining the Scope of the Problem

Bounding a problem is more than simply choosing which CAD models are appropriate for a particular analysis. A design engineer must be equally diligent about determining what he or she needs to know to produce a meaningful solution, realistic about what data they have vs. what they need, and be prepared to correctly interpret the solution so that a valid design decision can be made. The following tasks define the scope of the problem.

### 2.4.1 Defining the Goal

Any well-thought out analysis should be preceded by a well-defined goal. It is recommended that this goal be documented as one of the first sections of a project report. The goal should precisely state what outcomes are sought, under what conditions, and, when possible, how this data will be used. Scope creep, a problem in product development where marketing continues to add features to a design in progress, can also be a problem in setting up an analysis if the goal isn't clearly defined.

Examples of appropriate goals:

- Will stress in a given loading scenario exceed yield?
- Will we be able to reduce the first natural frequency below the operating speed of the motor?
- How many bolts, and of what size & grade are required to assemble these parts?

Examples of inappropriate goals:

- What will happen?
- Is it good enough?
- How will it be used?

### 2.4.2 What Questions Are Needed To Produce What Answers?

An important by-product of a well-defined goal is the physical outputs required from the analysis. If stress is a concern, which stress quantities are needed and on what parts? If deflection under temperature is an issue, how is that temperature distributed or how can it be determined experimentally or through preliminary analyses? In most cases, the primary question must be preceded by a series of discovery questions to ensure that all the data needed to make the answer to the primary question meaningful exists.

### 2.4.3 Which Parts Must Be Physically Modelled?

As will be discussed in the section on Boundary Conditions, users must decide what physical interactions contribute to the results of interest. The first step is to identify on which parts or assembly of parts a user expects to find the results of interest. If a user needs to know stress in a cast housing, that housing should be included in the analysis. The next step is to define which parts or environmental factors interact with the parts of interest and then determine, using the rules for defining boundary conditions, whether interacting parts need to be physically modelled or if they can be represented by loads or restraints. If the cast housing mentioned previously is bolted to another part, can that part and the bolts be modelled with a restraint or do they need to be explicitly included?

### 2.4.4 Asking the Right Questions

In many cases, an overly broad question might be a symptom of insufficient understanding of the limitations of the technology. If a user asks, "Will it break?" they may be expecting a straightforward, black or white answer. Unfortunately, the response to the question will be a field of stresses, displacements, or temperatures that, in themselves, don't provide any indication of success or failure. A more realistic question that can be answered directly by the system is "Will the Von Mises Stress in Part A exceed the safety factored yield strength under worst-case operational loading conditions?" Embodied in this question are the goal of the analysis and the allowable operating limits.

### 2.4.5 What Do the Answers Really Mean?

Over-confidence in analysis results is usually an indication that an analyst has unwittingly crossed the bounds of their problem-solving environment. If a problem's scope has been properly set with a well-defined goal and operational allowables, a user should be able to make immediate and meaningful decisions from the resulting data. However, a user should avoid trying to rationalize what good or bad means based on what they see after an analysis is run.

Additionally, a user should have a good feel for the validity of the results based on his or her qualification of the inputs. In most cases, a designer should be looking at stress or displacement levels with respect to stresses calculated using similar methods on previously established 'good' or 'bad' examples. Even full-time analysts rarely have enough data to make definitive judgements on stress results as compared to published allowables. This isn't a limitation of the analysis tool but acknowledgement of the fact that most inputs have a wide variety of possible values, or scatter. The output will never be more precise than the input.

Finally, users must remember that each analysis performed represents a snap-shot of one possible design configuration. One dimension or material property value is considered versus the total distribution of all possible dimensions or properties. To address this, a growing technology involves using a statistical variation of each input instead of a single value to derive a probability curve for each output of interest. This area of study is called probabilistic analysis. While not commonly included in designer level analysis tools, it is an important technology to watch for.

## 2.5 Chapter Summary

There are many factors that enter into the success of an analysis program. Clearly, an organization must determine that money and/or time is being saved over traditional testing. To achieve this, there needs to be a good level of confidence that the simulation mimics reality well. Many analysis projects start off on the wrong foot due to poor planning. Understanding the goals of the design project and the role of analysis in it can help make sure all subsequent work is meaningful and productive. Make sure the information you need is within the scope of the tools available and your ability to work with these tools. Finally, plan for the decisions you'll make with the data generated so that you'll know a better design when you see it or recognize the value of insight when it comes.

# 3. Basic Analysis Concepts

This chapter will introduce concepts that should be understood by all practitioners of FEA and will be discussed in more detail later in the book. Not all of these modelling topics will be directly applicable to the user of CAD-embedded analysis technology at the current state of the technology. However, with each release, more powerful and flexible capabilities, such as planar idealizations, are being added that will make this information relevant. Remember that as power and capabilities are added to a tool, the burden falls on the user to ensure their personal problem solving environment isn't violated.

## 3.1 Current State of the Technology

At the writing of this text, in late 2007, there are a variety of tools and approaches available to design engineers looking to incorporate analysis in their processes. Tools like COSMOSWorks from SolidWorks Corporation and Pro/MECHANICA from PTC allow users to access FEA functionality directly from the CAD interface and the modelling information is stored directly within the part or assembly data file. Other tools like DesignSpace from ANSYS, Inc. and "FEA in CAD" products from Algor, Inc. work alongside a CAD system transferring data back & forth between the CAD interface and an analysis-specific interface. Finally, the larger enterprise-wide systems like Catia from Dassault Systemes and NX from UGS (acquired by Siemens AG in May 2007) have scalable analysis solutions that vary in their capabilities and CAD-like integration.

Despite their differing approaches, each of these tools represents an unprecedented level of accessibility to FEA for design engineers. Design engineers can solve larger and more complex problems today than many full-time analysts could attempt 10 years ago. With this added accessibility comes the responsibility to understand both the inputs and outputs from an engineering perspective. The developers of many of the tools mentioned are working hard to build engineering guidance into their software to help users interpret the resultant data but it is inconceivable that these in-line advisors can anticipate every possible problem that users might encounter. Users are still responsible for the engineering content of their projects.

## 3.2 Engineering Prerequisites for FEA

As design analysis has evolved, the question of a minimum education level, job title, or years of mentored experience has been discussed in user forums and trade publications with no general consensus being reached. As mentioned in Chapter 1, the technology has, or will shortly become, a mainstream tool in the product design process. Such a condition renders the discussion of minimum requirements somewhat pointless. There is no watchdog group that is capable of enforcing these minimums. Engineering management, either intentionally or unintentionally, will allow designers and design engineers of all knowledge levels access to FEA without much control over usage. The technology is simply spreading too rapidly to police. For the foreseeable future, the burden will fall to individual practitioners to take the steps necessary to ensure they understand their bounds and limitations. When necessary, they should expand their personal problem solving environment. Otherwise, they'll need to work within it when expansion isn't possible. To that end, there are some guideline 'minimums' that users should consider before becoming overly reliant on their analytical work to make design decisions.

### 3.2.1 Engineering Mechanics and Failure Analysis

Chapter 5 will review the individual concepts in more detail but all users of FEA should understand the characteristics of the material they are using within the operating environments expected. They should understand when a material becomes nonlinear, transitions between ductile and brittle, and what factors impact the material's ability to perform its intended function. Users must also understand what failure mechanisms are likely to occur in their products or systems in all anticipated uses. Most users set up analyses and interpret results in terms of single static overload failures when buckling, resonance, impact, fatigue, and/or creep might be a more damaging failure mode.

As stated earlier in Chapter 2, FEA can't be counted on to replace engineering judgement. It just answers properly posed questions. If buckling is likely with the applied loads and restraints, a linear static analysis will not flag the user to run a buckling study and the results from the linear static study are not necessarily going to suggest how or at what load buckling might occur. A user must anticipate the potential response(s) of a system before he or she can predict that response. This requires some engineering insight into the behaviour of your systems.

### 3.2.2 Free-Body Diagrams

Free-body diagrams (FBDs) are one of the more basic skills taught to undergraduate engineering students yet still one of the more powerful ways to evaluate load distribution and macro level interactions between components in a system. A free-body diagram illustrates the sum of all forces interacting with a part or group of parts that respond as a single part. The FBD must account for magnitudes and orientations (vector quantities) of all forces and moments plus include any dynamic loads resulting from translational or rotational accelerations.

The steps for completing a general FBD are as follows:

1. Isolate the body of interest.
    - Typically the part or system being analyzed.
2. Choose a convenient coordinate system (CS).
    - Consider the benefits of a cylindrical or spherical CS where applicable.
    - Positive/Negative reference frames must be established for forces, moments, velocities and accelerations. Clockwise (CW) or Counter-clockwise (CCW) are commonly used references for rotational terms.
3. Place all external force vectors for that body on the FBD.
    - Use meaningful labels for these vectors as it is easy to lose track of references as the equations of motion are manipulated.
    - It is often convenient to break down vector forces (F) into their components based on the chosen CS; such as Fx, Fy, and Fz.
    - Don't forget to include a dead weight load acting at the center of gravity (CG) if applicable.
4. If there are translational accelerations or rotational velocities or accelerations, diagram the location of the center of gravity and any resulting dynamic loads.
    - Determine translational acceleration forces via the equation $F=Ma$; where M = mass and a = acceleration.
    - Determine rotational velocity (centrifugal) forces via the equation $F=MR\omega^2$; where M = mass; R = radius; and $\omega$ = rotational velocity.
    - Determine rotational acceleration generated torques via the equation $T=I\alpha$; where I is the mass moment of inertia for the body and $\alpha$ = the rotational acceleration.
5. Write out expressions that equate the sum of all the force and moment components to zero and solve for the unknowns in the system.
    - Remember that to solve a series of simultaneous equations, there needs to be as many equations as unknowns. If there are too many unknowns, there may be redundancies in the system, you may need to dig deeper into the problem or make assumptions about the unknowns.
    - It is often helpful to complete a FBD in a spreadsheet program or math software to track and manipulate variables and plot the sensitivity of one or more unknowns to changing inputs.
    - Computer-based dynamic or mechanism analysis tools can also be used as FBD generators for more complex systems or when multiple configurations must be studied.

Keep an eye out for redundant load paths in the system being studied. These are situations where reaction forces and/or load paths are dictated by the relative stiffness of the bodies in question. In these cases, the FBD may need to be solved in two or more steps to isolate the unknowns.

The results of a correctly constructed FBD can be used to define loads for an FEA model, validate results using reaction forces, determine the components that need to be included in the analysis, (especially when a redundancy is noted) and simply better understand the system in study. Users are encouraged to exercise their FBD skills on all applicable problems.

### 3.2.3 Stress-Displacement Calculations

On certain parts in some industries, the stress and deflection of interest can be solved to a reasonable level of confidence with closed-form calculations. Fortunately for most design engineers, there are plenty of references (most notably Roark – see section 13.4) that have a compilation of calculations pre-configured for a variety of geometries and load distributions. The most basic of these are the equations for uniaxial loading, pure bending, and pure torsion.

Another level of complexity is added when continuous bodies have small features with increased levels of local stress, or *stress concentrators.* Again, references available to design engineers[1] have compiled commonly occurring stress concentration factors to aid in the calculation of these localized results.

In many industries though, parts and/or load paths don't conform to these more straightforward stress scenarios and seem to defy attempts at closed-form calculations. It is for these problems that FEA is particularly well suited. Admittedly, FEA can be equally useful for solving the more simple systems although it might be considered overkill. However, the ability to see the load path in a system and identify areas of predictable response, (i.e. pure uniaxial, bending, or torsion) is valuable as a means to check FEA results on more complex systems. With so many variables going into even the most simple structural calculation, disciplining yourself to backup your analysis with any additional calculations you can is worthwhile.

Identifying applicable equations to represent all or part of a system's response also has indirect benefits in providing guidelines for the choice of solution types (linear vs. nonlinear, for example) and to provide insight into what problem parameters have the greatest effect on the response of interest. For example, if a region of pure bending can be identified, closed-form calculations can provide guidance on the relative benefits of increasing wall thickness or rib height simply by examining their effect on the bending moment of inertia.

---

[1] Peterson's Stress Concentration Factors (2nd Edition), Pilkey, W; John Wiley & Sons; 1997

## 3.3 Key Assumption Categories in FEA

Nearly every input to a finite element model is an assumption. A given FE model is, at best, a snap-shot of one possible configuration which a user hopes is representative of all possible configurations. When a user cannot draw this conclusion easily for any given input, he or she should explore the sensitivity of the results of interest to the variation of the inputs that couldn't be nailed down. The various inputs/assumptions required for most analyses can be categorized as Geometry, Properties, Interactions, Mesh, and Physics.

The validity of any analysis is dependent upon how well a user understands these assumptions. It is reasonable to say that if they were completely quantifiable, they would be 'knowns' not assumptions. Therefore, a finite element analyst needs to develop a realistic picture of the assumptions he or she is making including upper and lower bounds.

### 3.3.1 Geometry

When working with a CAD-embedded analysis tool, getting the geometry right may seem like the easiest part of the project. However, there are many aspects of CAD model preparation that can make or break an analysis, both from an efficiency and a validity standpoint. It is important to remember these points regarding the geometry used for analysis:

#### 3.3.1.1 Geometry is a Template for the Mesh

Despite the fact that much of the problem set-up is driven by the CAD model, the finite element solver still only understands "geometry" as the placement and shape of the nodes and elements in the mesh. Combinations of features that create small edges or narrow surfaces make it difficult for the mesher to build well-shaped elements in the model which can reduce accuracy and efficiency. These types of modelling techniques should be avoided. This will be discussed in more detail in the next chapter.

#### 3.3.1.2 CAD Geometry Can Only Be Representative

Due to tolerances that are specified and manufacturing variations that aren't necessarily controlled, nearly every part is as different from the next as two fingerprints. When assembly tolerances and variations are considered, it should be clear that the task of analyzing every possible combination of geometric variables approaches the impossible. Consequently, a design engineer must research and document the variation that may exist between the CAD model used for the analysis and the possible geometries in the field. At times, simply acknowledging that the geometry one is using for FEA is only representative of the as-manufactured part may be more important than attempting to quantify all the variation possible.

### 3.3.2 Properties

The properties assumption has two aspects which must be considered in the construction of a finite element model. The first is ***material properties***. This encompasses the input properties the solver requires to characterize the stiffness of the system in question as well as certain failure properties that might come into factor of safety calculations or mark the transition from elastic to plastic behaviour. This will be discussed in greater detail in Chapter 5. The second is the properties of the ***elements*** themselves. As a design engineer strays from a 3D solid element representation of a part into surface/shell or line/beam representations, element properties are required to completely define the physical properties of the element lost by the abstraction or idealization. The properties required for each type of element will be reviewed in greater detail in Chapter 6 as the needs of the mesh are discussed.

### 3.3.3 Mesh

Beyond the element properties, the mesh assumption covers the type, shape and size of the elements used in the model. As will be reviewed in Chapter 6, the accuracy (and often validity) of a finite element model will hinge upon the construction of an appropriate mesh for the problem at hand. Another factor that must be considered within the scope of the mesh assumption is making sure there are enough nodes and elements in areas of changing response, (such as in an area of high localized stress) to make sure the results of interest are properly captured. Systematically reducing the size of elements near high response gradients is called ***convergence***. This is handled with varying degrees of automation in the various FE tools available but in the end, it is the responsibility of the user to make sure the mesh is refined sufficiently in the right places.

### 3.3.4 Boundary Conditions

The Boundary Conditions (BCs) in an FE model represent the interactions of what you explicitly modelled with what you didn't. As advanced as today's analysis tools have become, the user is still responsible for deciding how to represent these interactions to the solver. BCs are usually defined in terms of loads and restraints. Techniques and options for BC definition will be discussed in greater detail in Chapter 7. While there are often multiple 'correct' ways to define BCs for any given problem, a general method for assessing the validity of a particular choice is to compare the resulting response from the analysis to the behaviour you would have seen had the actual components been inserted instead of the BC. Simply ask, "Can it really move like that?" This simple question has resolved many a boundary condition problem.

### 3.3.5 Physics

Each solution or study type available in FEA can only address a certain subset of the operation or physics a real-world system might encounter. Making conclusions about other physical behaviors not captured in that study type can be perilous. This can be one of the more difficult decisions for part-time FEA users to get their arms around. A linear static solution will be the workhorse for most of your analysis needs. However, many problems have nonlinear aspects that can invalidate decisions made off of a linear study. Similarly, failure due to vibration and resonance or buckling will not be predictable from a static solution. A design engineer using FEA needs to assess all the likely physical phenomena that might be required for the problem at hand and explore the variations in response, or more importantly, the conclusions based on that response, before whole-heartedly committing to a linear static course. Tips on identifying alternate physical models and their basic set-up requirements will be discussed in Chapter 8.

## 3.4 Assumption Sensitivity

As important as the assumptions themselves is the awareness that one is making an assumption. As stated previously, nearly every input to a simulation is an assumption. If you are not aware you are making an assumption when you apply a load or assign a property, then you aren't prepared to evaluate the effect that variations in that assumption have on the results of interest.

We make assumptions in FEA for a number of reasons. First and foremost, the inherent variability of most problem parameters force us to pick and choose which values will give us the most representative data with the fewest iterations. For many users, only one combination of parameters is ever studied. If the inputs didn't have one clear and unassailable representative value, this may not have been enough. Secondly, the limitations of the technology require users to model, approximate, or idealize a real world situation. Material properties are often not uniform or homogenous. Large sheet metal parts are rarely flat and assembled components are unlikely to be perfectly orthogonal. However, these are the most convenient way to construct and model the parts we are interested in. Lastly, try as we might, we may never know all the data we need to 'perfectly' model a problem. Plastic material properties vary as a result of so many things it is impossible to characterize them all. Outside the lab, a designer may never really know everything a consumer might do to their product. However, that same designer is still responsible for making sure all conceivable extremes have been covered. This is where understanding sensitivity comes into play.

In a nutshell, any change to one of the inputs for a model will have some effect on the output. That change may be trivial or significant. It may be linear and predictable or it may be a total surprise. Oftentimes, the change in the output does not necessarily affect the results of interest or the decision one might make from that data. The impact of a change on the design decisions a user makes due to a change in inputs is the *sensitivity* of that assumption. Of primary interest to the designer is the sensitivity of an assumption at its extremes. Things become problematic when the response at extremes straddles the boundary between acceptable and not acceptable. In this case, the result of interest could be said to be extremely sensitive to the assumption and prompt the designer to think through the problem more carefully. One of the first questions to be asked is if the offending extreme case was even physically possible. If it is, you may need to consider some testing to better understand this interaction or variable.

Since every input to an analysis is an assumption, it would be helpful for design engineers at all stages of learning in FEA to note the extremes of the key assumptions in geometry, properties, boundary conditions, meshing, and physics and consider the potential sensitivity issues that could arise. Preferably, this review will be included in the final project report with further discussion of any test models that were constructed to explore these issues.

## 3.5 Chapter Summary

Finite element analysis is an engineering tool and a certain degree of engineering background or understanding is important to properly setting up and interpreting a problem. Certain "building blocks" were discussed in this chapter but all users are encouraged to learn as much as they can about the materials, failure mechanisms, and alternate evaluation methods for their products before expecting to impact their process with FEA.

# 4. Material Properties

The stiffness of a part is primarily a function of geometry and material properties. Therefore, a designer must put as much thought into choosing material properties that represent the actual material as in the creation of geometry. Designers must also concern themselves with the material properties that allow prediction of part acceptability or failure. If you focus most of your analysis efforts on trend studies, getting the material properties close will most likely be sufficient. However, if you have hopes of making a reasonable prediction of pass/fail using the absolute stress and displacement results from your finite element analysis, knowing what properties are needed, what they represent, where they came from, and how they affect your results becomes very important.

## 4.1 Properties Required for a Simulation

As suggested in the previous paragraph, analysts need to consider two important sets of material properties:

1. Input Properties – The properties needed to calculate a response

2. Failure Properties – The properties needed to interpret the response

The material properties needed in either set will also be dependent upon the solution type you are using. Table 4-1summarizes the material needs of each solution (the material is assumed homogeneous and isotropic).

| | Input Properties | | | | | Failure Properties | | |
|---|---|---|---|---|---|---|---|---|
| Solution Type | Young's Modulus | Poisson's Ratio | Shear Modulus | Density | Coefficient of Thermal Expansion | Yield Strength | Tensile Strength | Compressive Strength |
| Static - No Acceleration or Temperature Loads | X | X | * | | | ** | ** | ** |
| Static - Acceleration Loads | X | X | * | X | | ** | ** | ** |
| Static - Temperature Loads | X | X | * | | X | ** | ** | ** |
| Frequency | X | X | * | X | | | | |
| Buckling | X | X | * | X | | | | |

*Table 4-1; Typical Material Properties per Study Type*

* Shear Modulus is not required if both Young's Modulus and Poisson's Ratio are specified.

** These failure properties are applicable to these solution types only yet may not be required depending on the goals of the study.

The material properties needed for a nonlinear material analysis get complicated very quickly and are beyond the scope of this document. If you are interested in exploring this further, you are encouraged to read the NAFEMS book,

"Introduction to Nonlinear Finite Element Analysis," (see section 13.2) an excellent and extremely readable primer on nonlinear concepts. The rest of this chapter will discuss the properties listed in the table above and provide insight as to their importance as part of your analysis project.

## 4.2 Stress-Strain Curves Basics

One of the most important tools in understanding material properties is the stress-strain (S-S) curve. All too often, newer users, especially ones without a strong engineering background, think of properties such as Young's Modulus or Poisson's Ratio as simply numbers needed to make an analysis run. However, you will be able to make better decisions based on your data once you understand the inter-relationships of all the properties.

An S-S curve is typically generated from a tensile test on a prepared sample of the target material. For metals, the most commonly used test standard is the ASTM E8 standard. It specifies a cylindrical sample that can be pulled at different strain rates (speeds) to best match the loading speed of the actual part operation. Figure 4-1 shows the actual tensile stress-strain data for A36 steel.

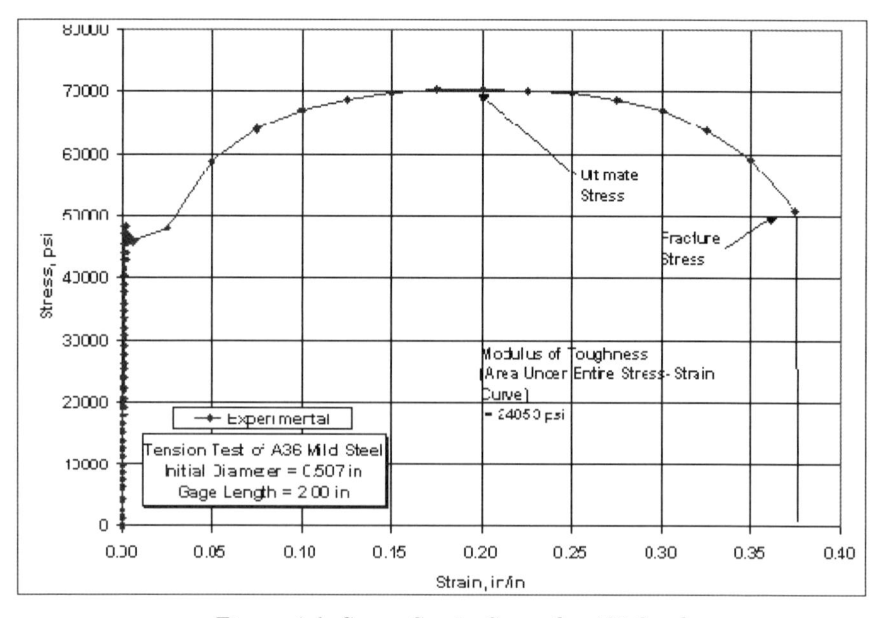


*Figure 4-1: Stress-Strain Curve for A36 Steel*

This curve represents the *engineering stress-strain* data. This type of curve bases the stress calculations (*stress = force/area*) on the original sample cross-section. As a sample is pulled, the cross-sectional area reduces due to Poisson's Ratio in the elastic region of the curve. As the material transitions into plasticity, where yielding or permanent set starts to occur, the cross-section reduces even more rapidly due to localized necking at the center of the sample. When the raw data is corrected for the actual cross-section at a given data point, it is called a *true stress-strain* curve.

From the curve, we can graphically determine many critical properties need for running an analysis and interpreting the results. Figure 4-2 identifies these properties for three different engineering metals.

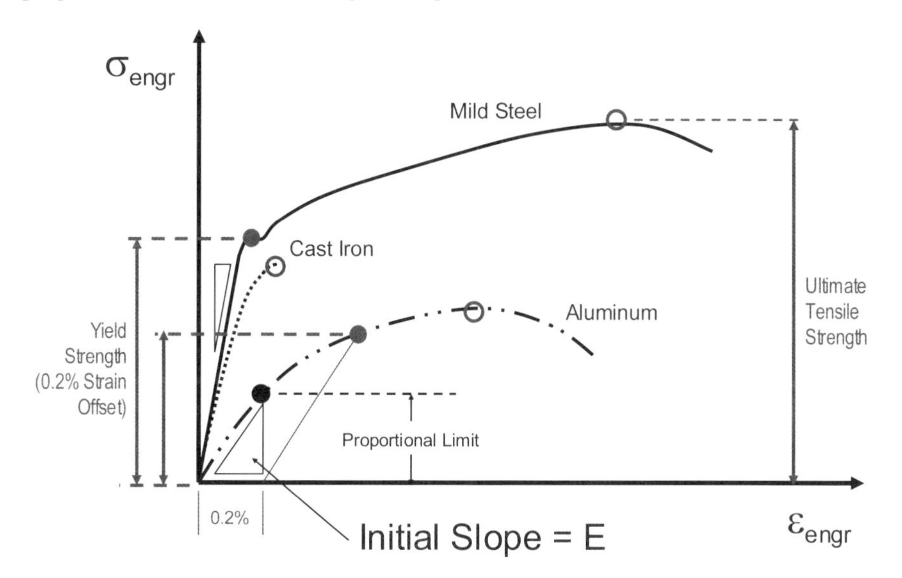


*Figure 4-2: Comparison of Different SS Curves*

For plastics, the most common tensile test standard is the ASTM D638 which requires a flat sample. Due to the wide variety of polymer compounds available, the S-S curve can take on three possible shapes as shown in Figure 4-3.

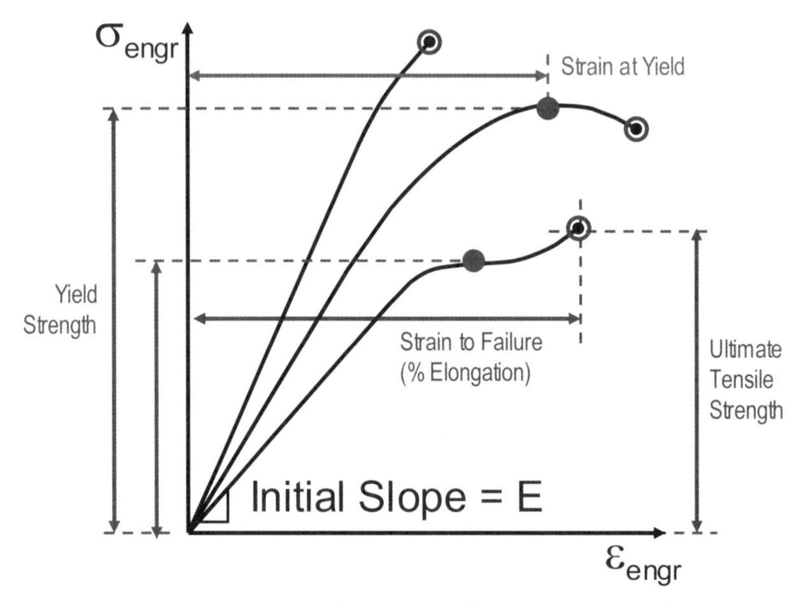


*Figure 4-3: Typical SS Curves for Plastic Materials*

Each of these properties will be discussed in more detail below but the descriptions will reference these figures where possible.

Each of the tensile test standards allows for testing at different strain rates and temperatures. This is because the properties that can be determined from these tests will vary due to temperature and speed. Figure 4-4 from GE Plastics shows how the S-S curve for their Noryl resin changes with temperature and strain rate.

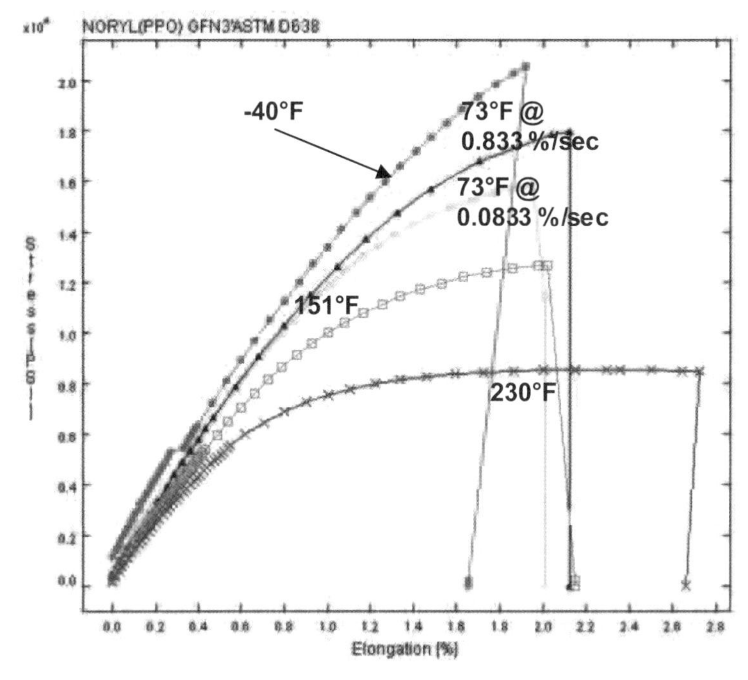


*Figure 4-4: SS Curves for GE Noryl*

Generally a part becomes stiffer, stronger, and more brittle at lower temperatures and higher strain rates. These three characteristics are important in selecting the appropriate material for a design and they can all be evaluated from the stress-strain curve.

Stiffness can be determined from the Young's Modulus of a material. The higher the Modulus, the stiffer the material.

Strength refers to the material's capacity to handle stress, either to resist yielding or fracture. The strength can be determined from the Tensile/Compressive Strength or the Yield Strength.

Toughness is an indicator of ductility. A material is tough if it can undergo large amounts of strain without fracture. The greater the strain at failure in a tensile test, the tougher a material is. Materials that fail at very low strains are said to be brittle.

Some comparative examples of S-S curves exhibiting these characteristics are shown in Figure 4-5a-e.

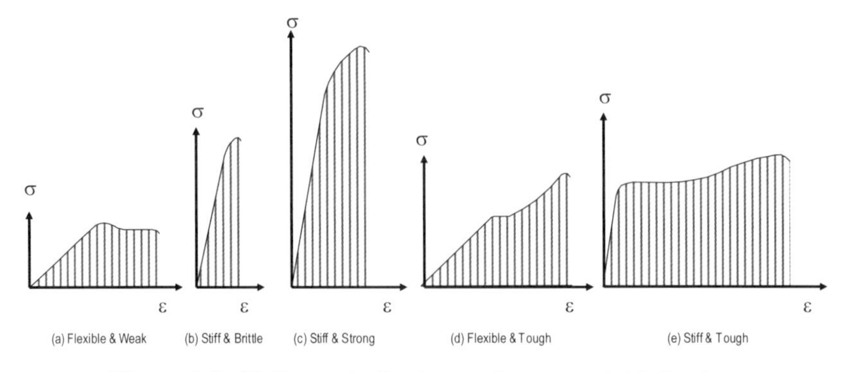


*Figure 4-5: SS Curves indicating various material behaviors*

It is recommended that you make an effort to get a representative S-S curve for your material tested at the closest strain rate and temperature to your expected operation. Even if you are simply planning to perform linear analyses, the curves can provide insight on the appropriateness of linearity as well as enhance your understanding of the materials you work with.

## 4.3 Input Properties

As stated previously, the input properties are those that allow your FEA code to calculate a response. In most cases, neglecting to specify a value for a required property will result in a solver error. In many cases, the input properties for families of materials vary only slightly, if at all, which makes sourcing the data relatively easy. For example, the Young's Modulus, Poisson's Ratio, and Density for 1020 & 4340 (and nearly all carbon steels) are the same. While any design engineer who uses steel regularly knows that the load carrying capacity of those two steels can differ by a factor of 2, the stress and deflection response to a given load of parts made with these steels will be identical. This knowledge should help many newer analysis users cut down on some iterations if an alternate material is a design option. By analyzing your part with generic steel and observing the resulting stress, you can choose a steel with the capacity to handle those stresses.

This isn't to downplay the importance of input properties. In fact, steel is somewhat unique in this manner. As you move into other metals such as brass, copper, or iron, determining input properties will start to get more difficult. Moving further into plastics or composite materials, input and failure properties become a field of study all to themselves. We'll touch on the reasons for this later in the chapter.

### 4.3.1 Young's Modulus

As mentioned previously, the Young's Modulus is the property that indicates a material's stiffness. It was shown in the S-S curve illustrations that it is the initial slope of that curve, or essentially the first measured stress value divided by the first measured strain value. It is typically represented by the letter, **E**. This is defined algebraically by Hooke's law which states that stress is proportional to strain via the Young's Modulus; $\sigma = E*\varepsilon$. Another common term for Young's Modulus is the Modulus of Elasticity although there are other elasticity moduli that come into play when a material has a nonlinear elastic range. For most steels, the shape of the S-S curve from start to the point yielding occurs is a straight line. These are truly linear elastic materials. Most other materials have some curvature in their S-S curves prior to the onset of plasticity. While stress is not linearly proportional to strain in these regions, if the load is removed before yielding occurs, the material will return to its original unstrained state. Other techniques for indicating the stiffness of a nonlinear elastic material at various non-zero strains are the Tangent Modulus and the Secant Modulus, shown in Figure 4-6.

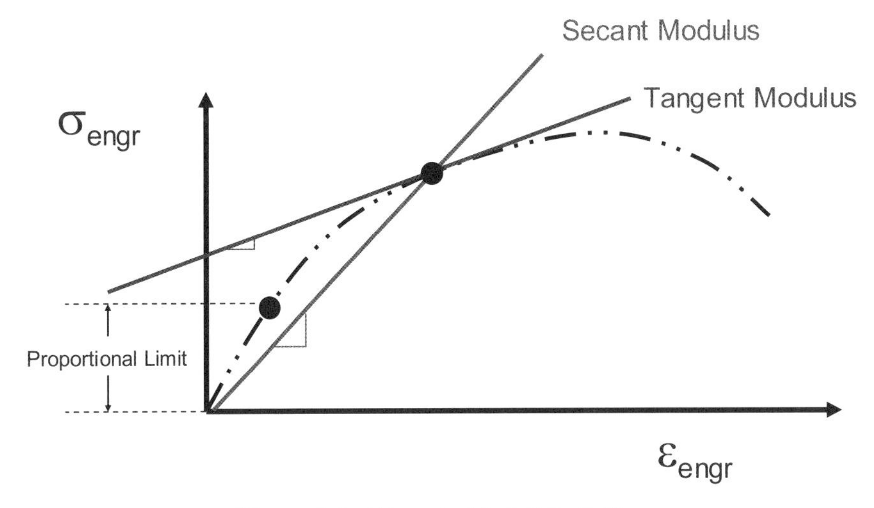


*Figure 4-6: Alternate Moduli Definitions*

The Tangent Modulus of elasticity is simply the local slope of the S-S curve at any given strain value. The Secant Modulus of Elasticity is the local stress divided by the local strain at any given strain value. These values can be used to bracket the stiffness response in a series of linear studies for a nonlinear elastic material when a true nonlinear solution isn't possible. Since the stiffness at higher strains can be much less than the Young's Modulus would indicate, these sorts of bracketing studies are important. Also note that until the stress levels reach the proportional limit, all three elastic moduli, Young's, Secant, and Tangent, are equal.

Young's Modulus is the primary material property used in the calculation of displacement. Under a given load, the higher the modulus, the lower the displacement. It is important to note however that under a given load, calculated stress is independent of Young's Modulus. Consequently, for a constant load and restraint scheme, you could change the material from plastic to aluminium to steel and the stress distribution and magnitude would never change. One way to help understand this is to look at some simple stress calculations with well-known solutions. Consider a simple rod in pure tension as shown in Figure 4-7.

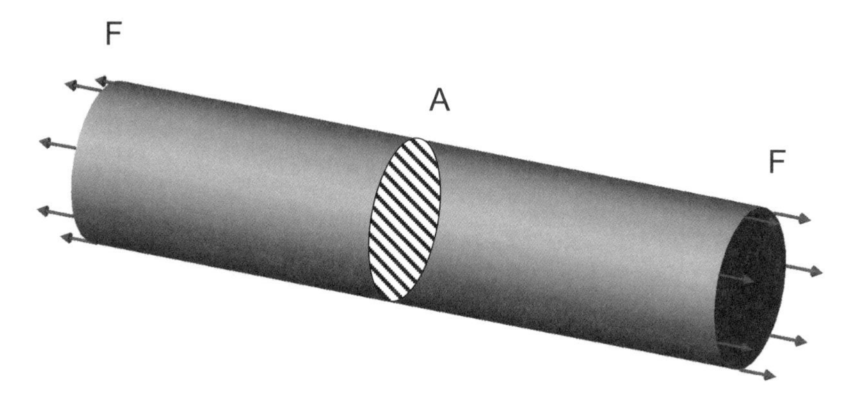


*Figure 4-7; Rod in pure tension*

The stress in this rod can be calculated using the equation, $\sigma = F / A$; where F is the axial force and A is the cross-sectional area. There are no material terms in this equation so it is clear the calculated stress will be the same no matter what the rod is made of. This holds true for more complicated geometries as well. It does not necessarily hold true for many nonlinear scenarios, especially when contact is involved. Nor does it hold true for all multiple material assemblies with redundant load paths or if the load in the model is applied through an enforced displacement.

The units of Young's Modulus are the same as stress, or force per length$^2$. In an inch-pound-second (IPS) unit system, Young's Modulus is specified as pounds per square inch (psi) and in the millimetre-Newton-second (mmNs) unit system, use MegaPascals (MPa)

### 4.3.2 Poisson's Ratio

Most of us have observed at some point in out lives that structures get thinner when you pull them and wider when you compress them. This trait is used quite often as a gag in cartoons as an unsuspecting character gets stretched accidentally and becomes a longer, thinner version of themselves. The property in materials that governs this stretch/thin behaviour is Poisson's Ratio. Literally, it is the ratio of lateral strain to longitudinal strain and is illustrated in Figure 4-8 and is typically represented by the Greek letter Nu or ν.

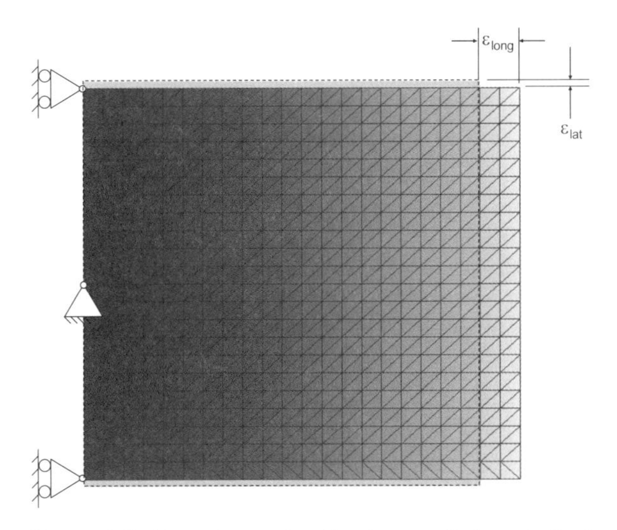


*Figure 4-8: Plate in tension illustrating Poisson's Ratio*

Note the restraints shown in schematic form in the figure. The middle of the back face is fixed in the vertical but the rest of the surface is allowed to slide freely as the part thins. This allows the plate to deform uniformly and naturally. Had the left face been restrained completely, or fixed, the thinning would have had to begin after the restraint as shown in Figure 4-9.

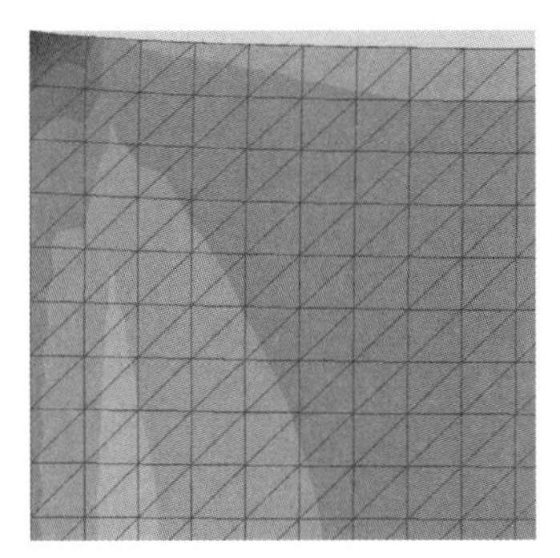

*Figure 4-9: Poisson effect stress on a plate in tension*

This is a Von Mises Stress plot and the dark spot in the upper left hand corner is an unrealistically high stress concentration caused by the shearing action at the restraint as the plate thins. These high stress areas or "hot spots" are often called Poisson Effect stresses resulting from the longitudinal strain caused by the Poisson's Ratio of the material.

In a 3D model, Poisson Ratio couples the strains in all directions. If the plate in the previous example was shown as a rod, as shown in Figure 4-10, you would see that the "lateral" strain refers to all directions normal to the applied load.

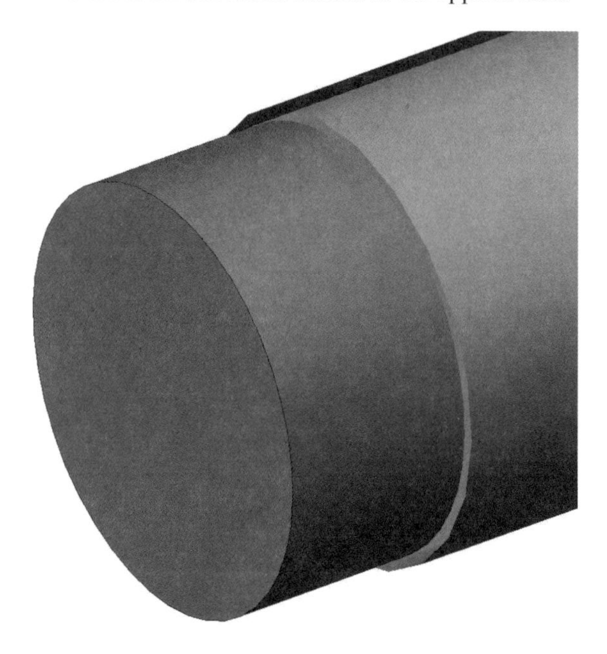

*Figure 4-10: Rod in Tension Illustrating Poisson's Ratio*

Poisson's Ratio is a unitless, or non-dimensional, value normally greater than 0 and less than 0.5 for most commonly used engineering materials. A material model with a Poisson's Ratio of 0 will solve in most FEA codes. A Poisson's Ratio of 0.5 creates a numerical error in most FEA codes and is typically only allowed for hyperelastic material models, used for modelling rubbers and elastomers.

In most cases, the value used for Poisson's Ratio will only have a small impact on the actual calculated results. Where the effects start to become noticeable is when there are significant compressive loads normal to each other in a system so they are working against Poisson's Ratio. Since this can be a difficult material property to source, it is common practice to assume 0.3 unless you find data to the contrary for your application.

### 4.3.3 Shear Modulus

Shear Modulus, G, is another elastic modulus and is related linearly to Young's Modulus and Poisson's Ratio through the equation,

$$E = 2G(\nu + 1).$$

In most cases, it will be easier to find Poisson's Ratio and Young's Modulus than Shear Modulus and most FE codes only require the specification of 2 of these properties. Some codes will ignore Shear Modulus if both are specified or prompt the user with a warning if the three values don't exactly work out via the previous equation. The error messages may range for a clear description of the property inconsistency to something more vague; such as "The material definition is not positive definite." It is worthwhile to understand how your software handles a violation of this relationship so you aren't caught unawares in the middle of a project. Consequently, if you know E and ν, leave the G field in your material input form blank.

### 4.3.4 Density

Density, often referred to by the Greek letter Rho, ρ, is the ratio of weight or mass to volume of a material. The higher the density, the more a given part will weigh. Most FEA codes require the specification of density in mass density terms, or mass per unit volume. However, many references provide density as weight density or even specific gravity. These terms can be converted using the following methods for the IPS unit system:

| Given | Divide by | To Get |
|---|---|---|
| Specific Gravity | 10,700 | Mass Density ($lb_f$-$sec^2/in^4$) |
| Weight Density ($lbf/in^3$) | 386.4 | Mass Density ($lb_f$-$sec^2/in^4$) |

*Table 4-2; Conversions for Typical Density Units*

### 4.3.5 Coefficient of Thermal Expansion

The Coefficient of Thermal Expansion is represented both by the acronym CTE and the Greek letter alpha, α. This is the property of a material that couples expansion or contraction to changes in temperature. Its units are in length/length/degree such that the larger a part is, the greater the size change for a given temperature change. The most important aspect of this property is that it reflects change, not absolute values. For example, if you are interested in the response of your system, normally stored at room temperature of 20C, to operation in a 100C environment, you would specify a $\Delta T$ of 80C, not a final temperature of 100C.

This property is used in structural analyses to determine stress or displacements resulting from temperature changes. It is not used in a thermal study, which calculates temperatures themselves. For a thermal analysis, you will need the properties, conductivity and specific heat. These are not required for a structural analysis.

## 4.4 Failure Properties

The second group of properties that a design analysis user must be familiar with are the failure properties of the materials in question. You can build the most detailed finite element model and research your loads carefully but in the end, you need to know what to make of the results. As stated previously, if your primary focus is on trend analysis, it may be more beneficial to know how analytical results compare to known performance. However, at some point in time, you will want to, or you'll be asked to, make a statement about part acceptability based on the stresses calculated. The more you understand the failure characteristics of your material, the better position you'll be in to make a confident prediction.

### 4.4.1 Ductile vs. Brittle Behavior

One of the most important characteristics of materials with structural requirements is the ductility, often termed toughness, in the application of interest. In general, materials are said to be ductile if they can undergo wide-scale yielding before fracturing. The opposite of ductile is brittle. One of the best indicators of ductility is the strain to failure on a stress-strain curve or Percent Elongation, which points to the same thing. One commonly accepted guideline is that Percent Elongation values greater than 5% are considered ductile. If it is less than that, you should anticipate brittle failure. In practice, if the Percent Elongation is near 5%, without any other data on how this material fails on similar parts in a similar environment, you should assume it could be ductile or brittle, evaluate it both ways and make a decision based on the more conservative result.

This distinction is important because materials that fail in a ductile manner will yield before fracturing. Consequently, the primary failure concern for ductile materials is yielding so calculated stresses should be compared to the Yield Strength. Von Mises Stress has been shown to be the most applicable stress quantity when evaluating ductile yielding.

Brittle materials will typically deform elastically until fracture. There will be no appreciable yielding so yield strength is not a good failure indicator. For brittle behavior, the Ultimate Tensile or Compressive Strength should be used to determine allowable stress. While Von Mises stress is a good predictor of yielding, it is not applicable to brittle fracture. Maximum and Minimum Principal stresses will tell you if a region of concern is tensile or compressive and what direction the stresses are oriented in. This is important since brittle materials tend to be stronger in compression than in tension. An even better failure indicator for brittle materials is the Coulomb-Mohr criterion. If you have access to this quantity in your post-processor, it has been shown to be a better predictor of fracture when compared to the tensile or compressive strengths of a material.

Remember also that ductility, as stated earlier, is environment dependent. A material that behaves in a ductile manner at room temperatures may fracture with no yielding at colder temperatures. Similarly, parts that exhibit good toughness in normal use may appear extremely brittle under shock loads such as might occur in a drop test. Manufacturing and other environmental factors can also affect ductility so you should try to understand the toughness of your material in the context of the expected use.

### 4.4.2 Choosing Appropriate Allowables

The failure properties we are about to discuss must be interpreted as guidelines rather than cast-in-stone absolutes. Since there are so many unknowns in the manufacturing, handling, and use of parts and in the simplification and construction of a finite element model, not to mention inherent variations in the materials being used, expecting a one to one correlation between analysis results and a datasheet yield or fracture strength is not realistic. Consequently, an important task before attempting any failure prediction is to put some serious thought into allowables. In the case of trend studies, the best way to determine your allowable stress is by analyzing parts that have test data associated with them and are of the same material with similar geometry and use. If you have access to developmental data showing multiple design iterations where a part that didn't work was fixed, analyzing the good and bad designs should highlight the stress levels subsequent designs need to live up to. If there isn't a clear distinction between the stresses in the bad and the good designs, then there is something in the system, the manufacturing, or the analysis set-up that you don't understand and you should reconcile this before moving ahead.

For parts or assemblies that can't be related to historical data, you'll need to base your allowables on the appropriate failure strength and apply a safety factor to those strengths. Some industries have codes or standards that outline safety factors and some companies have developed them over time. If you don't have the benefit of these pre-determined safety factors, you'll need to determine a safety factor that accounts for all the uncertainty in design, manufacturing, operation, material variability, and analysis idealization. This isn't always straightforward but is extremely important. It may take a few iterations to know if you are under or over-specifying the safety factor. Once an appropriate safety factor has been determined, it should be documented in your report with the supporting data or assumptions for future use.

### 4.4.3 Yield Strength

The Yield Strength of a material is the point where yielding or permanent set starts to occur. Once a material has crossed this threshold, the regions that have yielded will retain some strain when the load is removed. If the region that has gone plastic is small while the rest of the part remained elastic, the part may return to its original shape but the highly stressed spot will retain some residual stresses. When a part is overloaded in this manner only a few times, these residual stresses may never cause a problem. In a fatigue loading scenario, however, these are the spots where cracks are likely to occur first. Ideally, there should be no stress levels in excess of your safety factored yield strength for a more robust design.

The Yield Strength for most steels is very straightforward. If you refer back to the stress-strain curve in Figure-4-1, you can see a sharp transition at the end of the linear elastic region. This is the yield strength of this steel. However, if you look at the curves in Figure 4-2, you can see that for aluminium, there is no "knee" in the curve. For metals with no clear transition point, it is common to estimate the yield strength by drawing a line parallel to the initial linear portion of the S-S curve through 0.2% strain on the X axis. The yield strength is said to be the intersection of this line and the S-S curve. Note that this is an additional approximation on the material behavior and should be accounted for in your safety factor.

For plastic materials, ASTM D638 specifies that the yield strength is the stress at which the curve becomes horizontal (zero slope) the very first time. If you look at the S-S curves shown in Figure 4-3, you can see that this point is shape dependent. On the third curve, there is no point of zero slope. This curve represents a brittle plastic.

### 4.4.4 Ultimate Strength

The Ultimate Strength of a material is the maximum stress that material can endure. For most ductile materials, the ultimate strength is only slightly greater than the yield strength due to a relatively flat curve in the plastic region. This failure property is important when evaluating brittle failure. Remember that the Ultimate Tensile Strength, or simply, the Tensile Strength, of a brittle material is often lower than the Compressive Strength so you should make sure you are comparing tensile stresses to the Tensile Strength and compressive stress to Compressive Strength. Remember also that these failure strengths should be adjusted with a safety factor as well.

### 4.4.5 Fatigue Strength

The third most commonly used failure criterion is the Fatigue Strength of a material. This is essentially the maximum alternating stress a material can experience and operate for an "infinite" life. The Fatigue Strength is either published as a stand-alone value or can be derived from Stress-Life, or S-N, curves. The S-N curve for steels typically have a clear "knee" at a stress level called the Endurance Limit which indicates its Fatigue Strength, as shown in Figure 4-11. However, most other materials don't exhibit this characteristic so the Fatigue Strength is typically assumed to be the stress at which a sample survives $10^7$ or $10^8$ cycles.

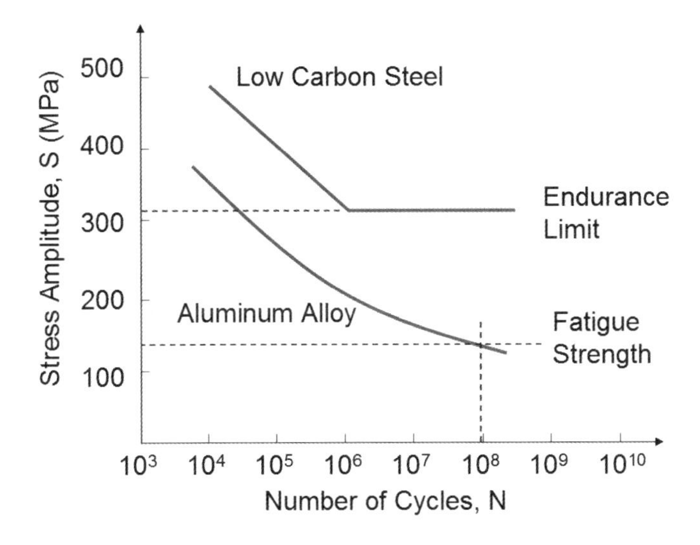


*Figure 4-11: Typical SN Curves for Steel and Aluminum*

Unless otherwise indicated, this data is measured from tests that subject a specimen to completely reversing loads. If the operation of your part doesn't subject it to completely reversing loads, with a mean stress of zero, then you'll need to adjust this fatigue strength to one that reflects the non-zero mean operation expected. This is typically done using a correction method such as the Modified Goodman correction. Details of this and other mean stress correction methods, such as Gerber and Soderberg, available in most mechanical engineering textbooks. There is an excellent overview of fatigue concepts in Collins[2]. Some design FEA codes now provide fatigue calculations based on the calculated results which greatly simplifies

---

[2] Collins, J. "Failure of Materials in Mechanical Design; Analysis Prediction, Prevention" John Wiley & Sons, New York, NY, 1993

the process. However, you are still responsible for specifying an appropriate Fatigue Strength and selecting the proper factor of safety. It is recommended that the minimum factor of safety you should choose in a fatigue environment is 2-to-1 and that is when you feel confident about all your inputs. This is due to the inherent variability in the fatigue failure mechanism itself. Trend studies, again, are a more reliable approach to durability investigations.

## 4.5 Guidelines for Commonly Used Materials

While every application is unique, there are some guidelines for certain material classes that are consistent and warrant discussion. In the end, however, you are responsible for understanding the materials you choose to the level your problem requires.

### 4.5.1 Steels

Steel is the best-behaved engineering material. As stated previously, the input properties for most steels are identical. A review of datasheets for various alloys may show the properties varying by a few percent. However, this shouldn't mislead you. Taking the approach that steel is steel when defining your input properties is reasonable. Consequently, the input properties for 'generic' steel are:

| Young's Modulus | Poisson's Ratio | Mass Density |
|---|---|---|
| 30e6 psi | 0.3 | 7.3e-4 $lb_f\text{-}sec^2/in^4$ |
| 200,000 MPa | 0.3 | 7.8e-9 $tonnes/mm^3$ |

*Table 4-3; Typical Steel Properties*

It is in the failure properties that steels are differentiated. A good approach for analyzing steel components in a linear material environment is to think in terms of 'generic' steel in the construction of the initial design model and then choose a steel that can handle the calculated stress levels. This will minimize over-specification of steel grades.

### 4.5.2 Grey Iron

Grey iron is a brittle material with a small, if any, linear portion on the S-S curve. Additionally, as with many brittle materials, the tensile properties are different than the compressive. Ideally, this material should be characterized by a nonlinear elastic material model that allows different responses in tension and compression. However, this is a complex input that isn't accessible to most design engineers.

Your best bet will be to review the tensile or compressive S-S curves for the iron you are using and choose a conservative elastic modulus if deflection is a concern. Also remember that the stress of interest is likely to be the Maximum Principal Stress, sometimes called P1 or Sigma1, and this should be compared to an allowable stress based on the Tensile Strength of the material since yielding will not happen. If your post-processor can report the factor of safety using the Coulomb-Mohr criterion, this is an even better approach.

### 4.5.3 Other Metals

Most other engineering metals have no distinct Yield Strength so the 0.2% offset Yield Strength will be your indicator of impending yield. The assumption of a linear material model should be reviewed for each new alloy. Some knowledge of the peak operating strains, calculated by an initial study, along with a representative stress-strain curve can help you determine if the linear solution is sufficient or if you'll need to adjust your understanding of the results based on the digression from the actual S-S curve. In many cases, maintaining operating stresses below 50%-75% of the Yield Strength will result in good linear approximation and is probably a safe design point if yielding is unacceptable.

### 4.5.4 Castings

Cast parts must be evaluated with the properties of their base material. Many casting grade alloys are more brittle than a machining grade alloy so you should review the Percent Elongation for the material you choose. Castings are also subject to variations in properties throughout the part. Depending on the process, it is not uncommon to have different strengths in smaller or thinner features than in the body of the part. Some directed research on your material and casting process is warranted if you are hoping to make better predictions on these parts. Also remember that castings are prone to voids and surface imperfections can accelerate fatigue. If nothing else, be prepared to increase safety factors on cast parts if you don't have extensive experience with similar parts made using the same process.

### 4.5.5 Weldments

Analyzing weldments may be one of the most poorly understood model types for design engineers. The welding process subjects the associated parts to an extreme material upheaval which results in geometry, property, and ductility changes. Assuming you can estimate what is really happening in a part local to a weld is dangerous. Even experienced practitioners of FE based failure analysis will refrain from drawing any conclusions from the stress levels reported in a model at a weld. It is still common practice to estimate the forces being carried by a weld joint in an FE model and then use standard weld tables and calculations to determine how much weld and what type can carry those loads. In a fatigue situation, where calculated stresses are important, it is still common to assume the stresses plotted directly at a weld joint in an FE model are suspect and to interpolate stress at a joint based on the surrounding stresses. If this is an important study type for your company, you are encouraged to research the latest methods for FE-based weld evaluation. For a better understanding of static weld sizing using FEA, readers are encouraged to read a paper by Weaver which explains FE-based weld sizing for shell models.[3] FE-based weld fatigue is a more complicated topic and references often differ on approaches. A few good references to start with are Dong[4] and Lotsberg[5]. However, Internet searches will yield more and more data as the topic becomes more mainstream.

### 4.5.6 Plastics

It has been said that the three most important things to know before attempting to analyze plastic parts are:

(1) Know the properties; (2) Know the properties; (3) Know the properties

The properties of plastics are extremely sensitive to temperature, processing, strain rate, environment, material orientation and other factors. Additionally, the S-S relationship of many plastics diverges from a linear assumption at small strain levels so a linear material model should be used with caution. Again, getting your hands on a representative stress-strain curve can help you understand the limitations of a linear assumption for your model. If you are tasked with making predictions for plastic parts based on the absolute stress and displacement results of your analysis (vs. trend studies), it is highly recommended that you develop a relationship with a testing firm that can provide you with enough data to characterize your material considering all the applicable factors which could impact your conclusions.

---

[3] Weaver, M.A. 1999 Determination of Weld Loads and Throat Requirements Using Finite Element Analysis with Shell Element Models - A comparison with Classical Analysis. *Welding Journal Research Supplement* Vol. 78, No. 4, p. 116s to 226s

[4] Dong P, et al. "A Mesh-Insensitive Structural Stress Procedure for Fatigue Evaluation of Welded Structures" International Institute of Welding, IIW Doc. XIII-1902-01/XV-1089-01, July, 2000

[5] Lotsberg, I "Fatigue Design of Plated Structures Using Finite Element Analysis" Ship and Offshore Structures 2006 Volume 1, Issue 1 Pages 45-54

## 4.6 Chapter Summary

Since much of structural FEA is performed to predict failure, understanding the properties of the materials being used is critical to achieving success. If you focus on trend studies, getting input properties close should be sufficient. However, the more you know about your materials, the better chance you'll have at making predictions on response which will allow you to push design envelopes further and innovate more aggressively. For a more complete discussion of the various material representations, or models, used in finite element analysis, readers are encouraged to review the NAFEMS document, "An Introduction to the Use of Material Models in FE." – see section 13.2.

# 5. Meshing

While many of today's design engineer focused analysis tools have managed to remove much of the pain associated with meshing, the process of fitting triangles or tetrahedrons into otherwise continuous geometry, the user still has the final responsibility for making sure he or she has the right mesh and a sufficiently refined mesh where it is required. Some understanding of the process and the components of a mesh can help you understand your options better.

## 5.1 What is a *Finite Element*?

A finite element is a mathematical representation of a simple shape that converts stiffness, forces and displacements into output quantities that are important to design engineers. There are many excellent references on finite element theory and element formulation if you'd like to research this topic further listed in the final chapter. To understand basic meshing concepts for the purposes of this book, consider the analogy of a spring.

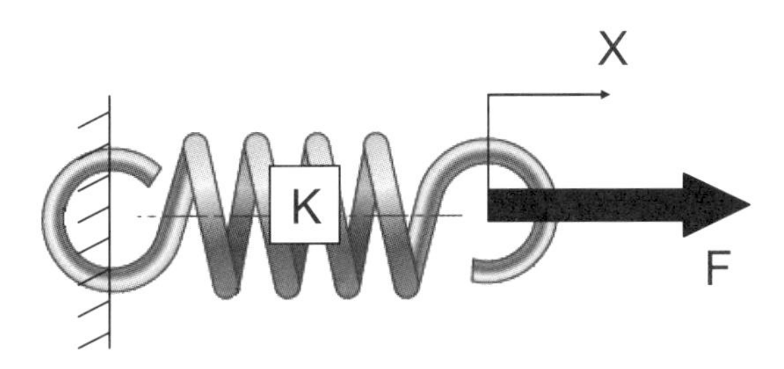


*Figure 5-1: Simple Spring Model*

In the system shown in Figure 5-1, the end deflection, X, applied force, F, and spring stiffness, K, are all related through the linear equation, $F = KX$. For a given force, there is a specific displacement. If the force is doubled, the displacement is doubled. However, practical experience with springs suggests that this relationship has a limit. If this spring is pulled too hard or too far, the coils start to yield and the spring no longer behaves as expected. In a finite element model, continuous geometry is represented by 'springs' in the form of elements. If an applied load forces one or more elements to deform beyond its limit, the response will also be unpredictable. Unfortunately, the error won't be as readily apparent as a yielded spring in the lab and is usually found only by meshing the same area with more elements that can share that displacement and comparing the results.

The actual nodal displacements are predicted by estimated *shape functions*. These are the mathematical functions that describe the potential displacements of each of the nodes. For the simple spring shown above (Figure 5-1), the location of the loaded end can be described as $F(x) = Lo \pm x$; where Lo is the initial spring length. For two dimensional elements, such as the triangular element in Figure 5-2, the displacements of the nodes would be functions of X & Y.

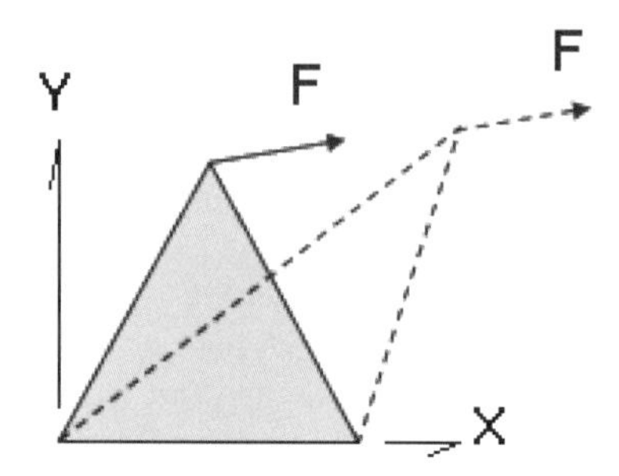


*Figure 5-2: Two Dimensional Triangular Element*

The final solution to a loaded system is the nodal locations as allowed by the combined shape functions of all the elements and nodes in the model where the work in each element, or spring, balances the applied loads and restraints. Once the nodal displacements are determined, the strain, change in shape over initial shape, can be calculated. Stress is then determined using Hooke's Law, $\sigma = \varepsilon E$.

You'll need to consider these three guidelines for obtaining a proper mesh.

1. The mesh representation of your CAD model must include all the features and curvature you feel are important to appropriately capture the desired response. Since the FE solver understands your geometry only by the placement of nodes and elements, the mesh of elements must approximate your geometry such that there aren't any glaring visual differences as you observe the model on the screen, at a minimum. Areas of tight curvature, spherical surfaces and small features should be examined and a more refined mesh in these areas added if required before kicking off the solution. Acceptable and unacceptable conformance is illustrated in Figure 5-3.

2. The initial mesh must also capture the global stiffness of your part or assembly. As stated previously, even though one element might capture a portion of a feature, if that element is asked to carry too much strain or shear, it will report back a falsely stiff response.

3. If a more accurate stress response is required, as it often is, the final point to consider is that, while the initial mesh might provide a good overall stiffness and global displacement response, an accurate stress in areas of rapidly changing stress may require a much finer mesh than you anticipate. They only way you'll know if you have enough mesh to capture the stresses well is to complete a convergence study. This will be discussed in more detail later.

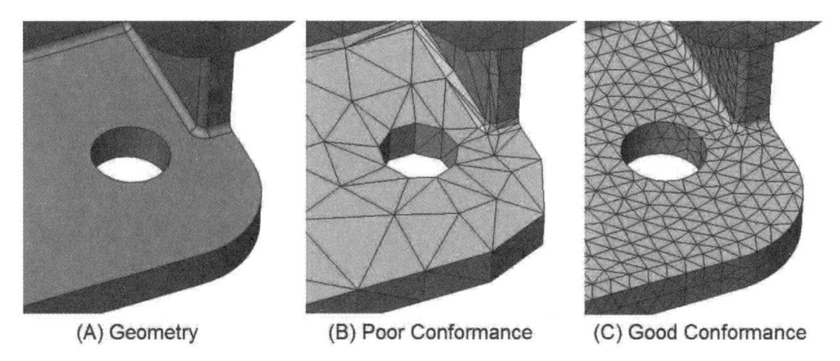


*Figure 5-3: Examples of Mesh - Geometry Conformance*

### 5.1.1 Shape

Finite elements come in three primary 'flavors'. These are line elements, shell elements, and solid elements. Each of these are best suited to particular geometry types. Understanding each of these types is important for designers even though your workhorse element will be the solid tetrahedron. Most popular design analysis tools provide options for all three so knowing when they might come in to play is worthwhile.

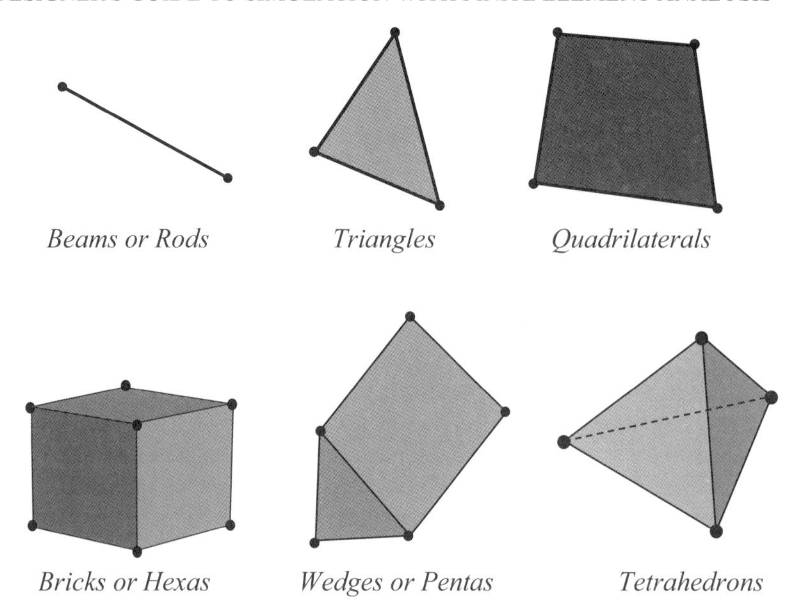


*Figure 5-4; Basic Element Types*

#### 5.1.1.1 Line Elements

Line elements represent parts that are long compared to their cross sectional dimensions. In other words, 2 of the 3 dimensions are much smaller than the third**Error! Reference source not found.** Figure 5-5a and 5-5b both show systems that are best represented by line elements.

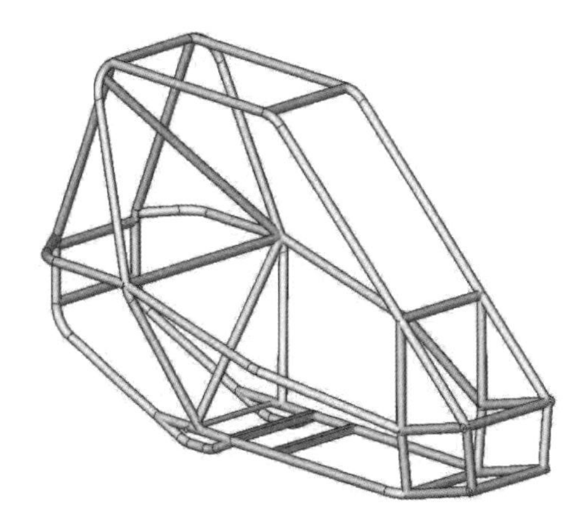

*Figure 5-5a: Beam model of dune buggy frame*

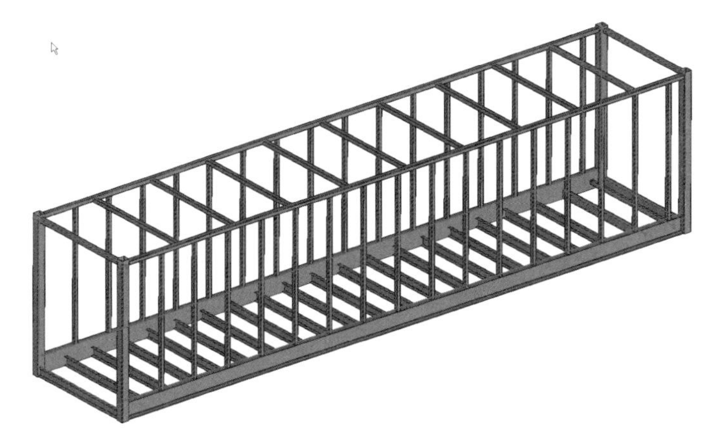

*Figure 5-5b: Beam model of railcar frame*

Line elements are represented physically by line, or wireframe, geometry. In addition to the line which represents the length of the member, the user must assign properties that define for the solver the geometry of the cross-section and the orientation of that cross-section on the line. The benefits of using line elements for members of this type is that they solve much faster than a solid representation of the same system and making changes to the section in pursuit of an improved or optimized design is extremely difficult.

#### 5.1.1.2 Shell Elements

Shell elements represent parts that are thin compared to their area dimensions. One dimension is much smaller than the other two. How much smaller this dimension needs to be in order to consider shells is a much debated question. If both area dimensions are 10 times the thickness, you are probably safe with shells. If the ratio is substantially smaller, shells may still be OK but you should try comparing some finely meshed solid models with the shell response of a simple part that is similar in geometry and loading. Figure 5-6 attempts to illustrate these differences.

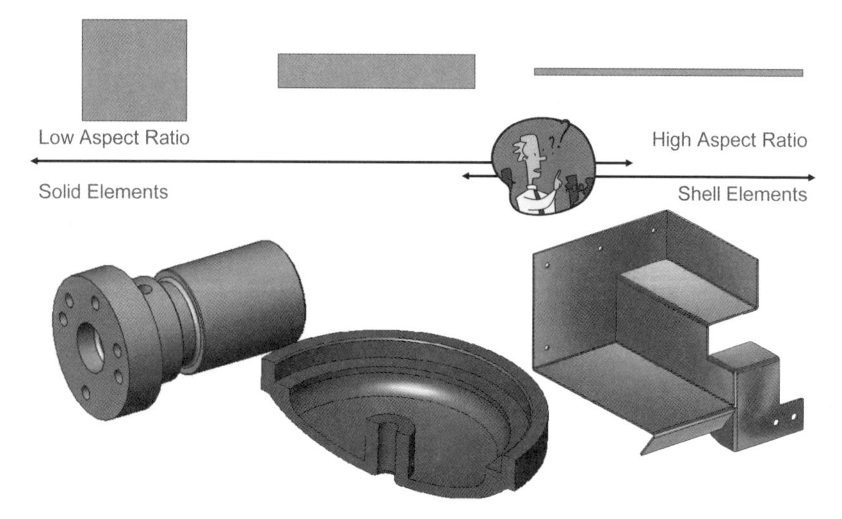


*Figure 5-6: Comparison of Geometry Aspect Ratio*

If the surface curvature is too tight the shell model may not be valid either. A ratio of curvature to thickness of 5 has been suggested but test models comparing solids and shells in a simple yet similar model is the best way to verify the applicability of the idealization.

Shell elements are either defined as four-sided elements called quadrilaterals, or *quads*, or three-sided triangular elements, commonly called *tris*.

Shell elements are physically represented by surfaces. The thickness of the surface is determined by a user-defined property. One of the great benefits of using shell elements when they are appropriate is that exploring the effect of increased or decreased thickness in a design doesn't require any CAD or geometry changes. The user just needs to type in a new thickness and re-run the solution.

#### 5.1.1.3 Solid Elements

Solid, or volumetric, elements essentially fill a defined volume. Unlike line or shell elements, no additional properties are required. Solid elements can be defined as six-sided hexahedrons, (called bricks or hexes), five-sided triangular prisms (called wedges or pentas), four-sided solids called tetrahedrons (often called tets) or square pyramids (which are useful for transitions between hex and tet meshed areas). The tet element lends itself most readily to automeshing arbitrary solids and is thus the element most familiar to designers. Bricks and wedges are capable of capturing response fields with fewer degrees of freedom than tets and are thus favored by analysts for compute intensive nonlinear or dynamic problems. However, the power and the flexibility of the tetrahedron have made it indispensable to analysis users at all skill and experience levels.

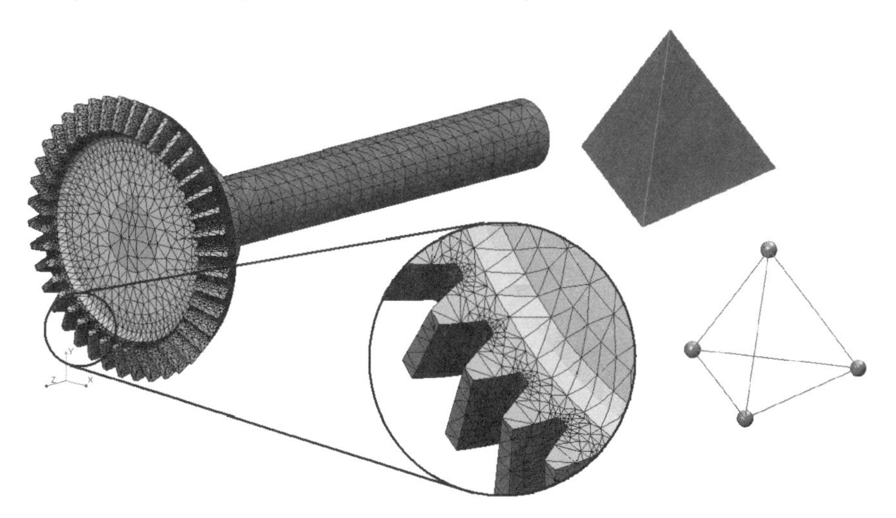

*Figure 5-7; Typical application of tetrahedral elements*

### 5.2 Convergence

A single element is only capable of capturing a certain amount of response. The more the results vary over a given area or volume, the more elements will be required to capture this variation. A mesh with too few elements in an area of rapidly varying response, such as stress, will still report an answer. However, that answer will not be the same as the answer with the optimum number of elements. The process of finding the optimum number of elements for important areas in the model is called ***convergence***. (Note that in nonlinear and other iterative analysis types, achieving numerical accuracy is also called convergence. These uses of the term should not get confused.) There are a number of techniques for achieving & checking convergence, both manual and automatic. Regardless of method, it is extremely important that all users, from designer to expert, check & pursue convergence diligently. An under-converged model that the user doesn't recognize as under-converged is one of the more common mistakes leading to erroneous conclusions since it usually under-predicts stress.

### 5.2.1 Element Order

An important property of an element is its order, generally meaning the order of the equation(s) relating nodal variables to variables inside the element or the mathematical complexity of the elements. The more complex an element is, the more response it can capture. Thus, a comparable level of accuracy can be achieved with fewer elements.

The simplest elements are *linear elements* which means that an edge connecting corner nodes is initially defined as a line and it can only get longer or shorter as the solution progresses. Going up one level in complexity, elements that can have edges defined as second-order equations are commonly called quadratic, parabolic, or even high quality elements. These element edges have a mid-side node, thus providing three points for a second-order equation to be defined through. Quadratic elements can be defined with initial curvature to better map to curved geometry. They can also deform into more complex shapes than linear elements thus capturing more stress or other response variation within a single element.

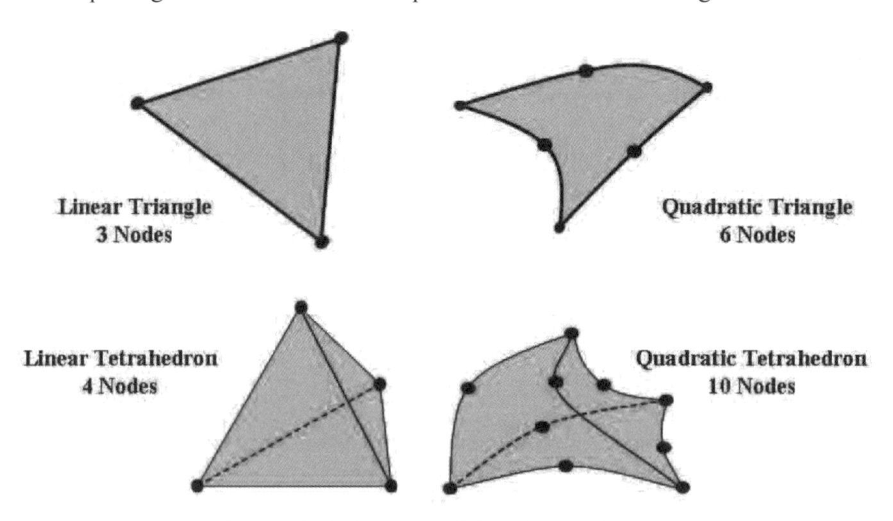


*Figure 5-8; Linear and Quadratic Elements*

For tetrahedrons, linear elements tend to respond overly stiff and in many cases do not produce the same results brick elements or quadratic tets might produce. However, if you are focusing on trend studies, a percentage improvement in stress or displacement from one study to the next with linear elements will still make for a valid comparison. Since linear tets (4 nodes) solve much faster than quadratic tets

(10 nodes), they still have a use in design analysis. If you are expecting to make design decisions based on the stress results from a single run, you should utilize quadratic elements in addition to proper convergence methods.

A different type of element technology, called P-elements, take advantage of even higher element definitions but are not available in all codes. These are discussed in more detail in Section 5.2.3. Readers who would like to learn more on element formulations and the various element types available in more complete FEA tools are encouraged to read the NAFEMS book, A Finite Element Primer.[6]

### 5.2.2 Methods for Checking Convergence

There are 3 primary methods for checking the convergence level of a finite element mesh once a solution has completed:

1. Relative convergence
2. Error estimates
3. Visual examination

Each of these has value although the order they are listed is indicative of the confidence you should have in each. Once understood, you should be able to determine how and when to use each.

Relative convergence checks, when performed manually, require you to compare the results from one solution to a subsequent one where the mesh has been systematically refined (smaller elements) either globally or in areas of concern. While these checks require more effort on your part, they are really the only way to know that you have the right mesh for the problem of interest. The downside is that this can become a burden when the model size, and consequently run time, gets large. This can also be troublesome when you had trouble getting the model to mesh at all the first time. Going back in & remeshing it in those cases probably doesn't sound too attractive. This is another reason why building FEA friendly CAD or spending a little time cleaning up a messy model is warranted versus trying to get creative with local mesh controls or using other tools that can force a fragile mesh on a difficult model. Despite the hassle, the first time you aggressively check convergence on a model where the stress levels changed from acceptable to unacceptable as the mesh was refined, you'll realize how important this is.

In the example shown in Figure 5-9, a machined part used in a high speed assembly system was meshed with three increasingly smaller element sizes in a potential area of concern. This feature was a small fillet in a machined component

---

[6] NAFEMS. *A Finite Element Primer*. Glasgow: NAFEMS, 1992

that showed a stress concentration under the applied cyclic loading. In the first model with a more coarse mesh, the stress in the small fillet shown was calculated to be below the fatigue strength of this steel. However, as the mesh was refined and the problem was solved again, the stress nearly doubled by the third iteration and was clearly in excess of fatigue allowables.

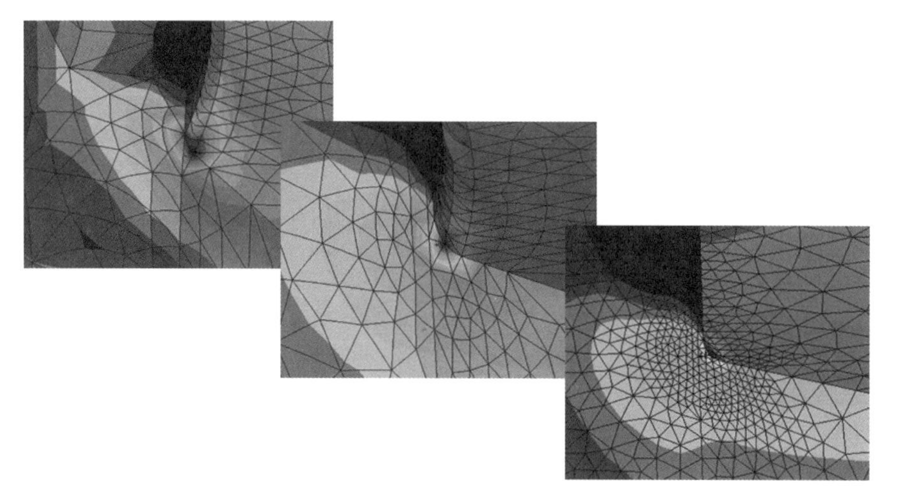

*Figure 5-9: Example of Mesh Convergence*

Another way to look at this problem, from a process standpoint, is with a convergence plot. This is essentially a graph where the x-axis represents the local mesh size, number of nodes, or even the solution iteration. The y-axis represents the quantity being converged, usually a stress component. These graphs don't need to be formal or precise. A sketched graph will suffice in many situations. A convergence plot for the previously mentioned example is show in Figure 5-10 with an estimate of where a fourth refinement data point might have landed.

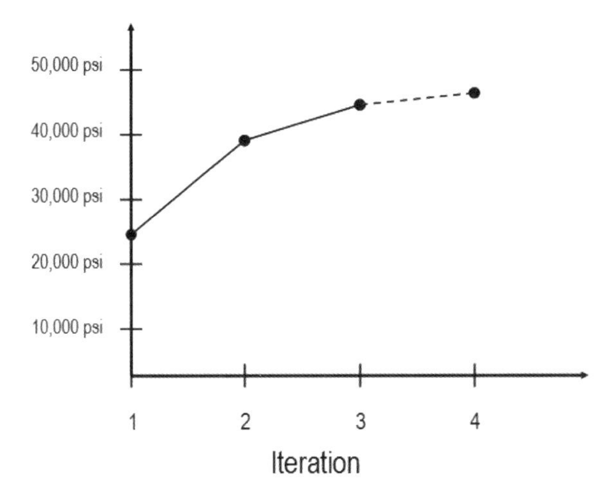


*Figure 5-10: Sample Convergence Curve*

From this plot, you can see that as the solution approaches the converged magnitude, the slope of successive line segments approaches zero. Noting the relative change in slope or the percent change in output from run to run is an excellent way to quantify the convergence of the model. In most cases, if stresses are changing less than 10% from one run to the next, you won't derive much benefit from pushing the mesh any further.

Another name for relative convergence is h-adaptive convergence. Historically, the symbol, *h*, has been used to represent the size, or height, of an element. Therefore, to achieve convergence in this manner the *h* of the elements in an area of concern is reduced until the stresses of interest stop changing appreciably. In fact, elements requiring convergence in this manner are called h-elements, which alone should suggest the importance evaluating convergence in your models. Most commercial FEA codes use h-elements as they lend themselves to automatic meshing well and are very computationally efficient.

Some analysis tools have implemented tools for automatically achieving h-element convergence called *h-adaptive algorithms*. When this feature is enabled, the software solves the initial mesh, identifies the areas in the model where a more refined mesh is likely to be needed & remeshes the part or assembly automatically. The new mesh is solved and the process is repeated until the pre-defined quantities of interest reach a converged level. Figure 5-11 shows the setup screen and Figure 5-12, successive meshes, from COSMOSWorks from SolidWorks Corporation.

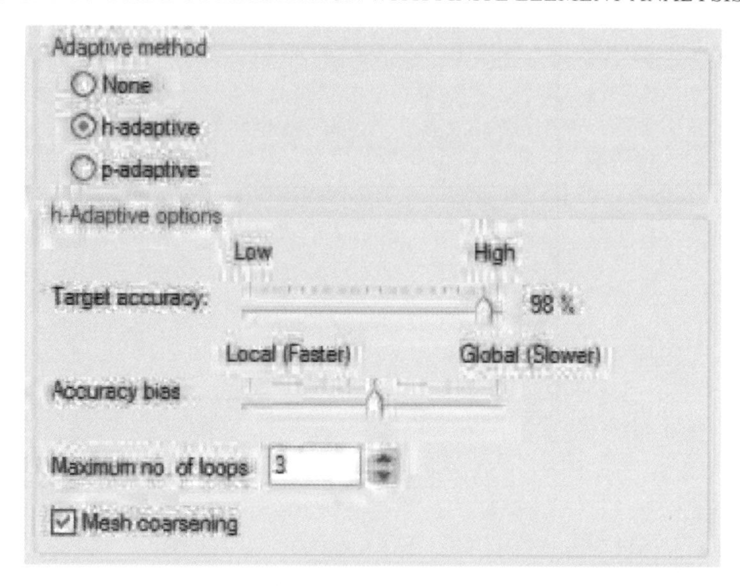


*Figure 5-11: Adaptive Options in COSMOSWorks*

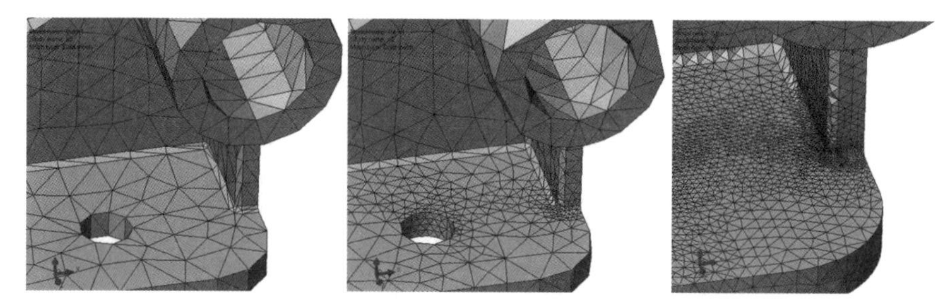

*Figure 5-12: Results From an H-Adaptive Convergence Study*

As can be seen from the h-adaptive set-up screen, this particular software package not only refines the mesh where required but also provides the option to have the algorithm coarsen the mesh where it doesn't need to be as fine as it originally was. Automatic h-adaptive algorithms are best suited to CAD driven analysis tools since the loads, restraints and properties are assigned to the underlying geometry, not the mesh itself. Therefore, re-meshing won't invalidate applied loads or restraints.

In addition to relative mesh convergence methods, users can also view a quantity, generically called an *error estimate*, to gain some feel for how well converged a model is. An error estimate calculation can be displayed in most analysis tools as a color contour plot showing nodal error instead of stress or displacement. Nodal error is most easily thought of as a measure of how rapidly a quantity is changing at each node in the model. Ideally, a mesh is fine enough such that, for any given node, all the elements attached to it have nearly the same stress. However, for more coarse meshes in rapidly changing areas of stress, adjacent elements, sharing one or more of the same nodes, may calculate dramatically different stress magnitudes. When adjacent elements have the same stress value, the local error would be zero. When they are different, the error is reported as positive. Most analysis post-processors also normalize the error in a mesh such that the highest error values reported are a product of the local error and the general stress magnitude. Large local stress variations aren't an issue at extremely low stress values but can mean the difference between deciding a part is acceptable or unacceptable in areas of high stress.

Another technique that can be used to evaluate the convergence level of a single solution is to compare the stress quantity and distribution with different nodal averaging methods. In short, a continuous fringe plot showing stresses reflects an interpolation between discrete values at each node in the model. As discussed above, the stress in adjacent elements will most likely vary by a large or small amount and, consequently, their estimate of stress at shared nodes may be different. It is most common for a post-processor to simply average the calculated stress at each node from the contributing elements for the color plots shown. However, some post-processors allow the display of unaveraged nodal results, possibly reflecting the maximum or minimum stress at each node. In a poorly converged model, a plot of the averaged stresses would be visibly different from a well converged model. If this option is available, it is a good check.

Figure 5-13 shows an error estimate plot for the model shown above.

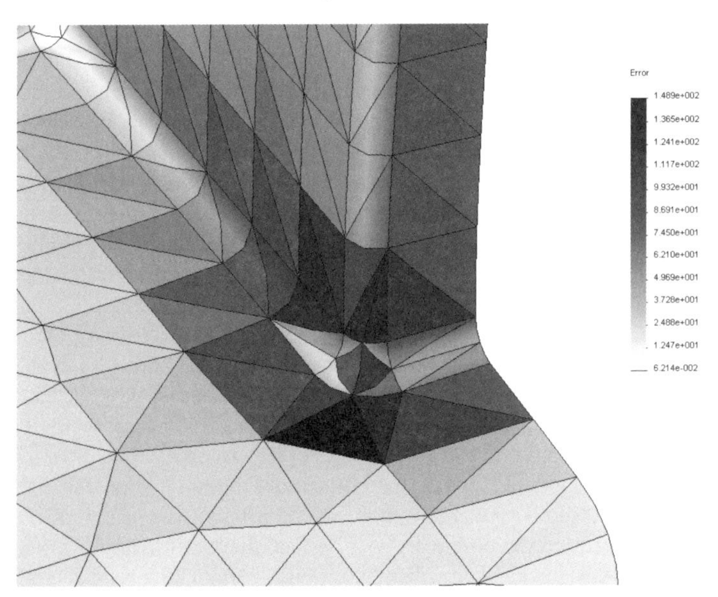


*Figure 5-13: Elemental Error Estimate Plot*

The last method for checking convergence is simply a visual examination of the results. In a part with smooth, continuous geometry, the stress contours should flow from element to element smoothly. When the stress contour lines have sharp, irregular boundaries, you should investigate the convergence quality in those areas. In Figure 5-14, the contours for a poorly converged mesh and a subsequently well-converged mesh are shown. You can clearly see the difference between the two plots. Finding the settings for an unaveraged stress plot may not be obvious in some designer level codes but certainly worth the effort. This is an important check to gain confidence in your results.

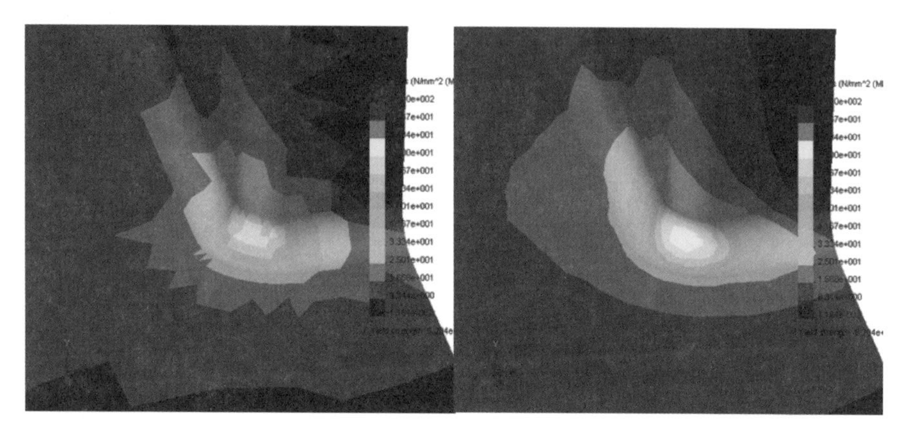

*Figure 5-14: Comparison of Elemental and Nodal Averaged Stress*

It is important to note that the default display for results plots in most FEA tools utilizes a continuous variation of color from one contour level to the next. This essentially smoothes over or blurs the variation in results across the model. It is very difficult to gauge the quality of the contour boundaries in this display. A better display method, also available in most analysis packages utilizes a hard, or discrete, boundary from one color range to the next so these edges are clearly delineated. The images shown previously in Figure 5-9 & Figure 5-14 use this discrete display mode. It is also helpful to make sure the element edges are showing for this examination because contour boundaries that follow element edges vs. flowing independently of them also indicate convergence problems.

You may have noticed in the preceding descriptions that, while the relative, h-adaptive convergence checking method involved time-consuming re-meshing and resolving steps, the error estimate and visual examination methods can be performed easily on any completed solution. This is why many part time and full time analysts will rely on the easier methods more heavily. However, the fact remains that the only true way to know how well converged your solution is is to use some sort of relative convergence check, either manually or automatically, if your software provides for it. The other two methods only give you a qualitative indication of convergence, not quantitative. Error estimates and visual checks are best utilized to compare the results on subsequent design iterations to a fully converged initial design solution. If you converge a model using adaptive methods then keep the mesh, both globally and locally, consistent for modifications to that design, you can check the contour quality and local nodal error to determine that the new design is similarly converged to the original one, making comparisons of stress valid. However, if your design changes are substantial such that the load path and/or local features change in topology, you may need to perform adaptive convergence on each iteration.

### 5.2.3 P-Adaptivity

One additional topic related to convergence bears some discussion. While most FEA tools in the industry today utilize h-element technology that requires h-adaptive convergence methods, some tools utilize an alternate called p-element technology. P-elements have a more complex mathematical definition than h-elements so they require more computational resources to solve. However, they have the added benefit of being able to capture larger changes in stress for a given element and most implementations of this technology allow the elements to essentially redefine themselves mathematically by changing the defining polynomial order to refine their local calculations without requiring a remesh. The benefit of this is an automatic convergence option called p-adaptivity. Most FEA codes are either h-element based or p-element based while some offer both options.

## 5.3 Mixing Element Types

One last topic with regards to meshing that warrants discussion is that of mixing element types. A detailed discussion of the ins and outs of mixing beam, shell, and/or solid elements in a single model is beyond the scope of this text. However, since more design engineer targeted tools are offering this option, it is important to know a few points to minimize the chances of misuse. When different element types are joined in a single mesh, there is a potential for a load discontinuity that may or may not be obvious when reviewing the results. This discontinuity primarily involves rotations. While different FEA tools support more complex combinations of nodal degrees of freedom (DOF), or load transferring capabilities, the most general implementation is:

- Nodes on a solid element – Forces in the three translational DOFs.
- Nodes on a shell element – Forces as on a solid and moments out of the plane of the shell.
- Nodes on a beam element – All six force and moment DOFs.

A beam element attached to a solid or shell element might behave as if the interface was a ball joint since the solid doesn't know what to do with the moments on the beam end. A shell attached to a solid element might behave as if on a hinge. In many cases, if the rotational load transfer isn't captured properly, as in Figure 5-15A, the solution will fail with an error indicating the model isn't sufficiently restrained. This is because the ball or hinge joint allows components to spin unhindered and the solver can't compute equilibrium. However, if the rotational load transfer between parts isn't captured properly but the system is otherwise geometrically stable, (Figure 5-15B), the model might solve without any error message but the results might be very wrong.

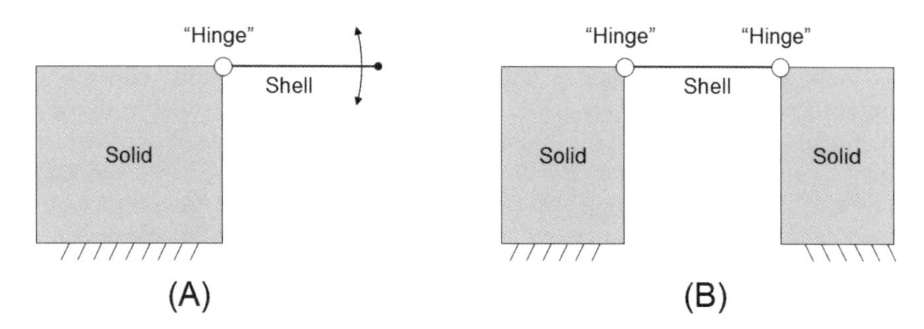


*Figure 5-15: Shell to Solid Interaction*

If you find yourself in a situation where you are considering using a mixed mesh for the first time, or for a new application, it is highly recommended that your review your modelling needs and your results with a more experienced analysis user so that you don't accidentally draw erroneous conclusions from the more complex model.

## 5.4 Chapter Summary

While meshing has become easier than it ever has been with today's tools, it is still the user's responsibility to make sure the right mesh is used and that it is up to the demands of the project underway. Try to take advantage of alternate idealizations and virtual structure within your analysis software to get to the answers as quickly as possible. As you explore alternate meshing options, make sure you understand any software-specific requirements for your particular tool. For a more detailed discussion of element types, mathematical formulations, and tips for selecting appropriate elements for mixed mesh models, refer to the NAFEMS Finite Element Primer.[7]

---

[7] NAFEMS, A Finite Element Primer, NAFEMS, Glasgow, Scotland, 1992

# 6. Boundary Conditions

Boundary Conditions (BCs) are essentially idealized representations, defined by loads or restraints, of the interactions between the parts you've modelled with the parts you didn't model. In theory, it's that simple. In practice, it may be one of the more difficult tasks in the construction of an FEA model. There are some guidelines that can help you make better boundary condition choices. In this chapter, these guidelines will be reviewed and the basic choices for loads and restraints will be defined. Finally, some tips for checking and evaluating your BC choices will be offered.

## 6.1 Building Blocks of Boundary Conditions

BCs can be generalized into three categories in most analyses:

1. Loads
2. Restraints
3. Contact conditions

Users of FEA at any level must be familiar with the capabilities and limitations of each. Many interactions can be reasonably modelled as any of these and, in many cases, there are numerous options within each category to consider. Guidelines for choosing the appropriate technique will be offered later in this chapter. However, first, you must understand each category and the various implementations of each.

### 6.1.1 Loads

Loads are typically nodal loads (forces), element loads (e.g. pressures or surface tractions), or body loads (e.g. gravity, acceleration, rotational velocity, or temperature change.) Loads impart no additional stiffness to a model so the geometry is free to deform naturally under the application of these boundary conditions. This is why some texts refer to them as *natural boundary conditions*.

In a geometry driven analysis tool, forces can be applied to solid or surface faces, edges, or vertices. Internally, to the FEA solver, these are redistributed as loads on the nodes associated with the selected geometric entity. They can be defined with a uniform distribution in all tools. However, some software allows for a non-uniform distribution where the load can vary in magnitude on the selected entity based on location referencing a user chosen coordinate system. This can be helpful to simulate forces that might be generated by increasing fluid depth in a tank (hydrostatic pressure) or contact with a flexible member or the load from solid particles such as coal or rocks.

In a nonlinear, large displacement analysis loads can be defined to change in direction as a structure deforms. However, in most analyses, loads are defined with an orientation and that direction is maintained throughout the calculation. This is an important limitation of loads. You need to determine if the load application remains oriented in the initial direction (linear load) or reorients itself as the geometry deforms (nonlinear aka “follower” load.)

The complete guidelines for a linear load are that the load must not change in magnitude, orientation, or distribution from application until the system reaches equilibrium. If you can’t assume this, you may need to consider a nonlinear solution. One nonlinear solution is using a contact condition to apply the load and this will be discussed shortly.

Pressure loads should only be applied to part faces in a 3D model. If you model your system using a 2D Plane Stress/Strain or Axisymmetric idealization, the pressure would be applied to an edge with the understanding that this edge represents an extruded or revolved face, respectively. Pressure loads act, at the solver level, on element faces and are always understood as a force per unit area input. Pressure loads can be applied uniformly or with a non-uniform, geometrically varying magnitude as in the force definition. A good example of this type of load is a deep tank where the fluid pressure varies with depth, typically the ‘Y’ dimension.

Body loads come in several varieties and act on the model volume instead of directly on the nodes and elements. Acceleration loads, such as gravity, translational or rotational acceleration, or rotational velocity act on the mass of the part such that areas of higher mass get higher loads according to the formula F = MA, where F is force, M is mass, and A is acceleration. Thermal loads, or temperature changes, act on the volume, not necessarily the mass, although the two are related. These loads are always expressed as a change in temperature since bodies in the real world always are at some temperature and our concern, as designers, is what happens when the temperature rises or falls. Cooler temperatures cause parts to shrink, as in a shrink fit assembly for shafting and warmer temperatures cause parts to expand as in an engine block.

In many cases, forces and pressures can be used interchangeably. The end result is the same if the total applied load is the same and the area of interest is remote from where the loads are applied. When acceleration body loads are used, it is critical that you check the mass and mass distribution in your model. This is especially important when shell or beam idealizations are used.

### 6.1.2 Restraints

Restraints are essentially “knowns” in the model since they eliminate degrees of freedom in the model, (the unknowns in the solution) and can be applied to faces, edges, or vertices. As "knowns", restraints are your way of telling the FEA solver that you know exactly what's happening at those entities. If you specify a zero displacement restraint, you are instructing the system to maintain the initial position for all points or nodes on that entity regardless of how much force is applied or how much deformation happens around it. Consequently, the responsibility falls on you to determine if you really know what will happen on those entities when the model loads up. One of the most common mistakes made by analysis users, at all levels, is to over-restrain a model by fixing or restricting movement when the actual system might see some local deformation. This is discussed in more detail in Section 6.2.

Unlike the physical world where gravity, contact and friction tend to keep things generally in equilibrium, the FEA world has no understanding of these factors unless you explicitly define them. More specifically, if you don't apply a restriction to the movement of a body, either in translation or rotation, the body is free to spin or float in that direction, or spatial degree of freedom.

In general, you'll need to apply enough restraints so that there is no possibility of unrestrained floating or spinning. This is not simply to produce meaningful results, although it would be difficult to evaluate deformation in a part that displaces through space an infinite distance. A fully restrained model is important to the mathematical engine behind every FEA system. An unrestrained direction, or degree of freedom becomes a numerical singularity in the calculations and prevents the system from determining the final state of equilibrium. Therefore, a well-thought out system of restraints that doesn't over restrain a system, or prevent natural deformation, is ideal.

In practice, contact conditions can often be used as a substitute for a restraint but there is an important difference. Restraints are always bi-directional so that if you restrain a flat face in its normal direction, that face can't move in either normal direction. However, contact conditions are uni-directional. A box sitting on a table can't translate downwards but if gravity were somehow to reverse momentarily, that box could float upward. In this case, the contact condition creates a state of pseudo-stability which is only mathematically solvable if the applied loading prevents, by the resultant sum of forces, movement in the unrestrained direction. This still creates a numerically ill-conditioned model which some FE solvers can resolve and others might have trouble with. One common solution is to apply a soft spring that attaches one or more points in your model to ground in the directions that are not restrained or restricted by contact. If the stiffness of this spring is low enough, it will have no impact on that desired response of the system but serves to improve the numerical stability of the model. Many FEA systems provide an option to engage soft springs automatically if an unrestrained degree of freedom is encountered. Care should still be taken to ensure that this option is used to stabilize directions you can't or shouldn't restrain versus directions you don't know how to or forgot to restrain. They shouldn't be used to correct sloppy modelling since the results could be unpredictable if employed carelessly.

By default, a restraint is a zero-displacement, positional restriction. Consequently, since calculated displacements in FEA report the distance displaced from the original shape, a restrained entity will report zero displacement in the final results. However, you have the option in most analysis tools to apply what is termed, enforced displacement. These are restraints at a position other than the initial, zero position. Applying an enforced displacement to a face will force that face to a new location and restrain it there. At that point, all the other restrictions of a restraint apply. This is an important technique when, as mentioned previously, you know exactly what will be happening at that location. A common misuse of enforced displacements is to displace a face or edge when the actual deformation has a bending or rotating component to it. In the example shown in Figure 6-1, the cylindrical journal at the end of a cantilevered member is expected to displace upward a certain distance as it interacts with a rising jack stand.

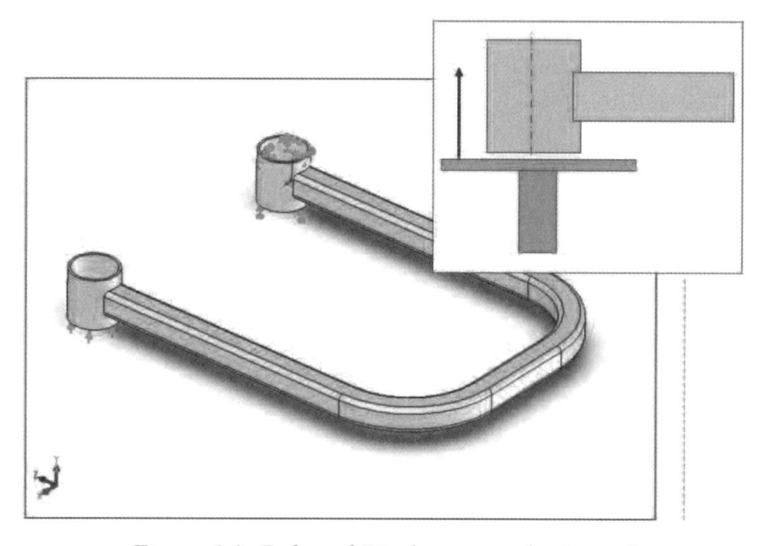

*Figure 6-1: Enforced Displacement of a Journal*

While the top face of the jack stand could be assumed to move in pure translation, remaining horizontal, applying that same enforced translation to the bottom of the cylinder, while initially in complete contact with the stand face, would negate the rotations expected by the bending of the arm. This will result in a fictitious applied moment on the journal to maintain the prescribed horizontal orientation. The moment imposed by the enforced displacement caused unreasonably high stresses to be calculated in the arm compared with a more natural contact loading. At first glance, the incorrect solution may seem valid but could lead to grossly erroneous design conclusions.

Restraints, in a CAD-embedded system, are typically applied to geometric entities and the directions they restrain are relative to that selected geometry. Table 6-1 suggests some ways to take advantage of this geometric associativity:

| Geometry | Components Restrained | Behaves Like | Other Restrictions |
|---|---|---|---|
| Flat Face | All | Perfectly bonded to an infinitely rigid part | This is a pretty significant restriction by itself |
| | Face Normal | Sliding Surface; "Hockey puck on ice" | No portion of face can lift off of surface. Other restraints or contact conditions required to stop this sliding. |
| | Face Normal and One In-Plane | Body on rollers or wheels that can translate in one direction only | No portion of face can lift off of rollers or initial position. No possibility of sliding and/or compression/expansion in the direction normal to the expected rolling direction. Other restraints or contact conditions required to stop the body in the sliding direction. |
| Cylindrical Face-Hole | All | Perfectly bonded to an extremely rigid pin | Full circumferential "bond" forces hole to remain in-place and perfectly round. No "egg shaping" due to radial loads. |
| | Radial | Hole or Journal in a stiff portion of a part on a rigid pin | Part can slide axially or spin on pin but axis is fixed and hole must remain perfectly round. No "egg shaping" due to radial loads. Other restraints or contact conditions required elsewhere on the system to prevent sliding or spinning. |
| | Radial & Axial | Hole or Journal in a stiff portion of a part on a rigid pin held tightly by a clip or nut in a frictionless interface; no slop. | Part can spin on pin but axis is fixed and hole must remain perfectly round. No "egg shaping" due to radial loads. Other restraints or contact conditions required elsewhere on the system to prevent spinning. |
| Cylindrical Face-Boss | All | Perfectly bonded within an extremely rigid hole. | Full circumferential "bond" forces pin to remain in-place and perfectly round. No "crushing" due to radial loads. |
| | Radial | A rigid pin in a zero-clearance rigid hole. | Part can slide axially or spin in hole but axis is fixed and pin must remain perfectly round. No "crushing" due to radial loads. Other restraints or contact conditions required elsewhere on the system to prevent sliding or spinning. |
| | Radial & Axial | A rigid pin in a zero-clearance rigid hole. held tightly by a clip or nut in a frictionless interface; no slop. | Part can spin in hole but axis is fixed and hole must remain perfectly round. No "crushing" due to radial loads. Other restraints or contact conditions required elsewhere on the system to prevent spinning. |
| Spherical Face | All | Perfectly bonded to an infinitely rigid part. | This is a pretty significant restriction by itself |
| | Radial | Zero-clearance Ball Joint in a stiff portion of a part attached to a rigid base or fixture | Part can revolve in 2 axes on ball but the center of the ball must remain fixed. The spherical face must remain perfectly spherical with no "egg shaping" or "crushing" due to radial loads. Other restraints or contact conditions required elsewhere on the system to prevent spinning. |

*Table 6-1: Common restraint techniques*

An important note about this table is that these combinations of geometry and restrained components represent the most common selections used in design. Taking this one step further, if you find yourself selecting a combination other than the ones listed here, you should be sure you understand the ramifications for deviating from more commonly used restraints. Take the time to understand the physical analogy, as in the "Behaves like..." column before proceeding.

Restraints can often be specified in a coordinate system reference frame as an alternative to a geometric reference frame. Many codes allow selection of Cartesian, Cylindrical, and Spherical coordinate systems for the creation of restraints. In these cases, you should verify that the axes of the coordinate systems you wish to use are properly oriented. Small deviations in orientations can cause unexpected results. Use this option with care.

### 6.1.3 Contact Conditions

Contact conditions define interactions between parts in a model. Hence, they wouldn't normally fall into the realm of Boundary Conditions since they act within the boundary of your model. However, as defining this once difficult analysis condition becomes easier and the speed at which these problems can be solved ceases to be a hindrance, analysts at all levels are forced to re-examine long held beliefs about introducing contact into a model.

Historically, some of the most creative load and restraint schemes were developed to avoid having to utilize nonlinear contact conditions, or actual part-part interactions, in a finite element model. In the early 1990's, applying a contact condition required meticulous mapping of nodes on one contacting face to the other so that node-node contact elements called Gap Elements could be defined. These elements were essentially conditional springs of sufficient stiffness that turned on when contacting faces approached each other, thus preventing penetration, and turned off when these faces retreated, allowing lift off. Choosing a high enough stiffness to minimize penetration while not making the springs so stiff that the contact faces chattered numerically, was part science, part art. Another limitation of these early contact elements was that they couldn't calculate relative sliding between the contacting parts. The consequences of excessive sliding ranged from erroneous results to a model that wouldn't solve. Even when the models did solve correctly, including contact took a model into the nonlinear realm where it would take a prohibitive amount of time to complete.

The tools of today allow designers to simply indicate that inter-part contact is to be included in the model and the software defines the appropriate parameters and mesh automatically. Large amounts of sliding, as much as practically required in most cases, are routinely allowed, even in the most basic systems and the solution times are tolerable for most design schedules. Contact 'stiffnesses' are derived from local body parameters and require little or no user involvement, thus removing the last hurdle to mainstream use of part-part interactions. With these advancements, choosing part-to-part contact has become a viable alternative to questionable applications of loads and restraints. In fact, for most users of FEA in product design, a case could be made to say that contact conditions should be the first thought for interactions near areas or bodies of interest and should be replaced with loads or restraints only when the substitution is obvious or unavoidable, as in the case of fluid pressure in a tank.

In most tools, two contact options are offered; No Penetration or Bonded contact. No Penetration contact is the traditional contact type where bodies are allowed to impact, slide against and lift off from each other in a more natural response to the applied loads. This does require a nonlinear solution and has been implemented in various codes with a small displacement formulation, large displacement formulation, or both. If appreciable sliding, contacting surface re-orientation, or surface conditions switch from compression to separation, a large displacement formulation should be considered. Subtle yet significant error can occur if these effects aren't properly handled. Examples of this are discussed in Section 8.1.3.

## 6.2 Guidelines for Determining the Model Boundary

Before choosing the best way to model the interactions between what you modelled and what you didn't, you must first decide where to stop modelling. The extent of the components included in the model is called your model boundary. Once established, the idealized interactions at this boundary are your boundary conditions.

Your choice of BCs for any given interaction must impart or allow behaviour at that interaction that the assembly or environmental conditions they represent would have imparted or allowed. For instance, in Figure 6-2, the flat bottom of a filled container is resting on a table in its unloaded state. If you chose to only model the container, that part would represent your boundary and the interaction with the table would need to be a boundary condition. An obvious option to represent this interaction might be a restraint on the flat bottom, fixing movement in the vertical direction and thus supporting it as it would seem the table does.

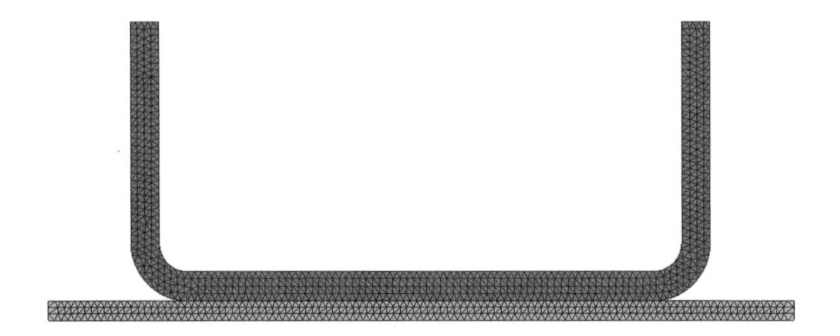

*Figure 6-2: Mesh of Container on Table*

However, if the loading in the actual, physical container caused the sides to bend outward, the resulting moment on the corners could force the bottom to bow up in the center. Since the restraints forced the bottom to remain flat, as if it was glued to the table, you might draw an incorrect conclusion about the deformation or stress in the container. The images in Figure 6-3 & Figure 6-4 illustrate this difference.

*Figure 6-3: Displacement of Container with Restrained Bottom*

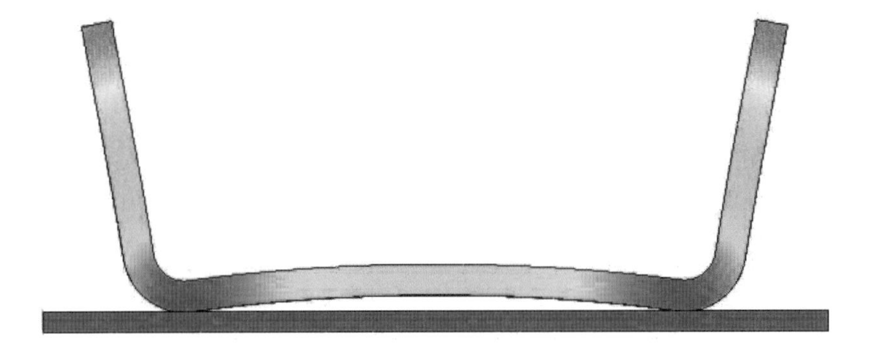

*Figure 6-4: Displacement of Container with Contact at Bottom*

Since the assumed restraint does not allow the response the actual interaction would have allowed, it is reasonable, in this case, to expand the boundary of the model to at least include the table top. Why 'at least'? If the table itself was flexible such that it might bow under the weight of the container, that flexure could additionally alter the response on the part of interest. If using restraints to model the table legs / table top interaction didn't allow a natural flexure, then the legs might need to be included in the model as well. As you can see, there is potential for a snow ball effect where you might be tempted to simply model every conceivable part in your system and all the systems it interacts with to ensure the proper response has been captured. A more systematic approach, however, is warranted. Two such approaches will be compared.

The first can best be described as the *sledge hammer approach.* In this, the temptation to model all conceivable parts is indulged but in a controlled process. You should start by identifying the part or parts that you need or want results on. Identify the features where interactions occur, take your best shot at idealizing those interactions with loads and restraints using the information presented in this chapter, and solve the model. Document the results of interest. Then build another layer of parts into your model, essentially expanding your model boundary. Use bonded or contact conditions where appropriate. Identify the new boundary interactions, apply the most reasonable loads and restraints you can establish and solve the model again. Repeat this process until the results of interest stop changing from run to run. At the point where adding parts no longer improves the response, you have reached the most efficient model boundary. In most cases, this can be reached in 1-2 iterations. Use common sense to determine which parts to include and how detailed these parts need to be. Supporting parts, on which you have no interest in detailed results, can often be greatly simplified as long as the main stiffness and geometric characteristics are maintained. With practice, you may be able to short-cut some of these iterations. That's where the second approach comes into play.

A more thoughtful approach to boundary determination requires you to make some judgments about the relative stiffness of interacting parts. In general, if the part of interest is much stiffer than the part it interacts with, then the interaction cannot control the local deformation, as a restraint might, so a load is appropriate. A good example of this is fluid, air or water, in a container. While the fluid can impart a load on the container walls, the walls will deform of their own accord and the fluid has to go along for the ride. Conversely, if the part of interest is much less stiff than the interacting part and there is no sliding at the interface forcing the part of interest to conform to the contours of the stiffer part, then you may be able to anticipate the deformation caused by the stiffer part and a restraint would be appropriate. An example of this might be the welded interface of a sheet metal cover to a cast or otherwise beefy housing. **The absence of sliding and/or conformance is an important qualifier.** Consider the compression of an elastomer seal into the o-ring groove of a stationary housing. There is no question that the stiffness differential is great. However, the unrestrained conformance of the seal is the result of interest so a restraint would not be appropriate since a restraint will force all selected surfaces to maintain their initial shapes.

In a more subtle example, consider the connecting rod and pin shown in Figure 6-5.

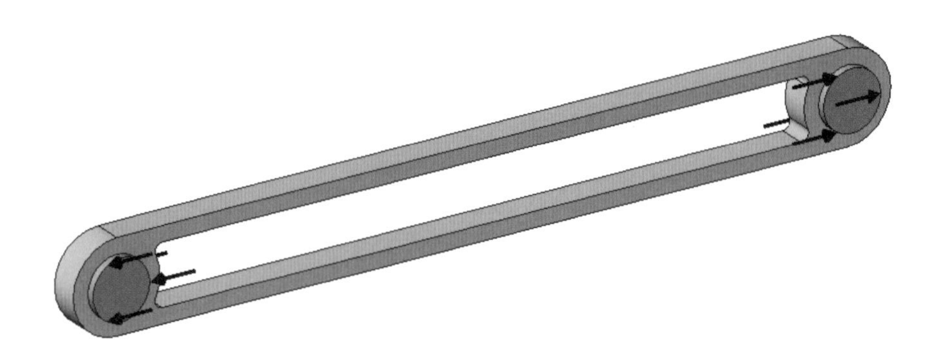

*Figure 6-5: Connecting Rod in Tension*

The thin walls of the connecting rod suggest that it will be much less stiff than the short pins. If the radial clearance is negligible, it might be reasonable to assume a restraint for the interaction between the connecting rod and each pin as the part is pulled. Figure 6-6 A – C shows one end of the rod modelled three different ways.

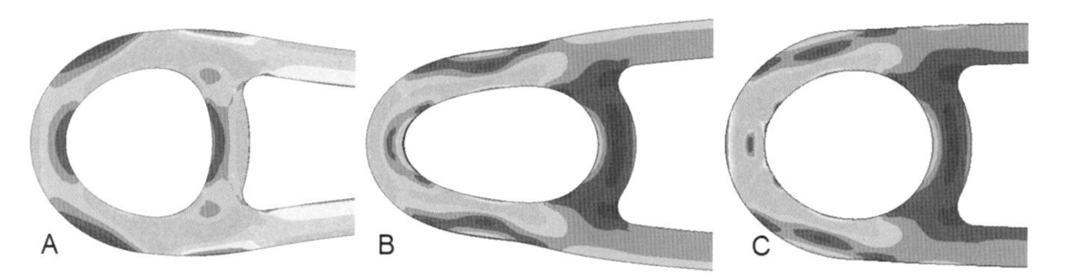


*Figure 6-6: Different End Results for Connecting Rod in Tension*

In Case A, a uniformly distributed load was applied to the entire circumference of the hole. Thus it pushed on the outside but also pulled on the inside. The resulting shape of the hole reflects this. In Case B, a Bearing Load distribution was defined which applies a greater magnitude of force on the face in the direction of the resultant vector and then tapers the load off around the circumference to zero, much as the contact pressure between a shaft and hole would create. This however does not force a cylindrical shape at the end of the rod which a stiff shaft would have so a greater degree of "egg shaping" is observed over Case C, where a contact condition between the two parts was modelled, including tangential sliding between the parts. While the results from Cases B & C are similar, Case B does not show all of the potentially critical stress areas identified by the contact model, Case C.

Based on these simple examples, you can see that your choice of model extents does make a difference. Interestingly enough, with a casual examination of any of the examples presented, you could probably convince yourself that they were all correct. It is only when multiple methods are compared that the differences become obvious. Other techniques for evaluating the validity of your choices will be expanded on in the next section.

## 6.3 Checking Boundary Conditions

As was alluded to in the beginning of this chapter, and hopefully supported by the subsequent discussion, properly indicating model boundary interactions requires as much intuition as science. Before you get disheartened, remember that it isn't magic and practicing engineers and designers are regularly successful. The real key is to be open-minded about what is really happening at your interactions, brainstorm all the options available to you and be prepared to try it 2-3 ways to ensure that the interaction responds as you expect it would. The better you understand the capabilities and limitations of your software, the better you'll be able to identify when an interaction requires special handling. You may not always know the best way to model an interaction but you should at least be able to recognize you've exceeded your knowledge in that area and go for help.

The most important tool you have for confirming your boundary condition choices or recognizing a problematic area is a rigorous checking process. Most load and restraint mistakes can be identified by using simple visual and query based techniques. Again, you may not know how to fix a problem once you see it but at least you'll see it. That's over half the battle.

### 6.3.1 Review of Input Sensitivity

One of the greatest tools for evaluating BC choices is to review the sensitivity of the output of interest to the load or restraint technique you tried. If you were able to identify several ways you might pursue loads and restraints, try each and see if the results of interest change appreciably. If they don't, then the results are not sensitive to this BC scheme. This should give you a good level of confidence that the decisions you make based on these results will not be affected adversely by the interactions modelled. If they do differ enough to suggest different conclusions from the data, you need to explore the differences to determine if one of these methods gives you correct or better results or if another option might be required.

### 6.3.2 Non FE-Speak Discussion with Colleague

Another technique that can produce unexpected benefits is to review your model setup and assumptions with a colleague who isn't familiar with your model or, even better, with analysis in general. Explain your choice of load, restraint, or contact in plain terms including what you expect this BC to do for you and why it is a valid representation of reality. You'll be surprised at how often this simple act of communication can bring to light some interesting contradictions between what you planned and what you actually modelled. It is critical that when this colleague tells you something doesn't make sense, you don't respond by saying, "Trust me, this is how you do it." If you can't make that person understand, you may have taken a leap of faith that isn't grounded in reality.

### 6.3.3 Summation of Loads

One of the hardest loading mistakes to catch is a mis-entered load magnitude. In today's tools, units can be entered on the fly resulting in models with dimensions in millimetres, forces in Pounds, pressures in Pascals, and any combination thereof. A mistake in unit specification will not be readily apparent using the visual feedback in the pre-processor and will be tough to see in a scan of the input definitions. If you are comfortable with multiple unit sets in a single model, you may have anaesthetized yourself to the shock more traditional analysts might get if they saw Pounds and Newtons in the same model. Additionally, certain tools might apply a force over a set of multiple faces as a total distributed load whereas other codes might apply the indicated force magnitude to each face. The end result can be quite different but the visual feedback in the pre-processor will be the same. A third case where load magnitude can be misapplied is in pressures across composite faces. In moderately complex solid models, interior faces on pumps, valves, and pressure vessels might have numerous faces of various sizes. In your selection of faces to apply a fluid pressure load, it is commonplace to inadvertently miss one or two. The symbols on the screen typically indicate the perimeter of a selected face so a missed face bounded by selected faces will appear selected. However the force imbalance caused by this omission can lead to erroneously high stresses.

For these reasons, it is highly advisable to estimate the expected resultant load on a model using free-body calculations as discussed previously or other methods more applicable to your particular problem. Then, either in the model definition stage, if your code allows it, or immediately following the solution, check to see what the software thinks the resultant load is on the model. If you expected your system to be force balanced, (i.e. sum of all applied forces and moments is zero) and the system thought you had a measurable force component in one or more directions, your results shouldn't be trusted and you should track down the error immediately.

Techniques to accomplish this vary with the different software packages. Look for options to sum all the applied loads or even selected groups of loads as a pre-run check. After the solution has completed, check force summations calculated by the solver. Check reaction forces at restraints for another way to evaluate loads. Some software packages even allow you to query model forces in a free-body diagram environment. Make sure you check moments as well in case the magnitudes are correct but accidentally off center.

## 6.4 "Unrestrained" Models

One last area of this deep and broad topic that will get mentioned is the class of problems that can best be characterized as 'unrestrained". These are systems such as:

1. Components in a mechanism that translate in space as they are loaded
2. Hand-held devices that have no true "fixed" frame of reference
3. Bodies in flight supported by a balance of forces
4. Systems supported by comparatively soft & flexible members

In these cases, finding that abrupt change in stiffness that suggests likely interactions for restraints isn't possible. Special consideration is required when modeling a part that has no physical restriction to deformation since any restraint is likely to over-stiffen the part.

The most common technique for handling these systems is to apply a perfectly balanced set of loads and/or pressures such that a summation of forces, as discussed previously, yields no resultant force or moment. While this may properly define the physical interactions, recall that your FEA solver requires a path to ground in all spatial degrees of freedom. Therefore, a common technique is to specify soft springs in the model that remove translation and rotation without inhibiting the strain caused by the loading. Choosing points to apply these springs can be tricky since they will essentially define your points of zero displacement. If your point selections are ideal, you could even apply point restraints in the proper directions without adversely affecting your model. However, springs give you an added cushion of safety. If they don't restrict deformation but simply anchor the part in space, the stress solution will be valid and the displacement solution will depend on your anchor points.

A common technique to check your point restraints or springs is to apply only a temperature change body load to your model with the planned anchor points in place. This load should cause stress free expansion if there are no restraints on deformation. If you observe stresses of meaningful size at the anchor points or the forces in your springs or restraints are above round-off error, you may want to look at alternate placement for your restraints.

Some tools provide a "soft spring" option in the solver that identifies unrestrained degrees of freedom and applies a soft spring to every node of the model, thus anchoring it in space. This can be a fast and easy way to resolve an unrestrained model or one where restraint is due to contact conditions. However, you are cautioned not to overly depend on this option to facilitate sloppy modeling. Make sure you've applied all the possible restraint to control the response of your model before falling back on global soft springs or you may get unexpected results. Don't forget that incorrect results often look temptingly like valid results.

Another technique available in many codes is called "inertial relief". While springs provide a stiffness based path to ground, inertial relief applies acceleration to the model to balance any residual force based on the equation Force = Mass x Acceleration. This technique was developed and is still intended to provide equilibrium for structures that are in a state of "moving equilibrium" such as an aircraft. At a constant speed, the system has balanced forces yet no physical restraint. Inertial relief provides a path to the more applicable dynamic balancing that keeps the aircraft aloft. For most land-based systems, soft springs will provide a better solution in unrestrained or under-restrained systems.

## 6.5 Chapter Summary

Thoughtful selection of Boundary Conditions may be one of the most challenging aspects of analysis. In many cases, the difference between acceptable and unacceptable is very small from a setup standpoint and it is easy to discount the impact this difference might have on the results. Since no FEA tool automates this decision for you, taking time to explore the alternatives available will help you make better decisions with more confidence.

# 7. CAD Model Construction

The convergence of CAD (Computer-Aided Design) and FEA over the past years has been a primary catalyst for the growth of FEA in product design. The promise has been there since the late 1980's but didn't become practical until the late 1990's with the growth of COSMOSWorks from SRAC (now SolidWorks Corporation), DesignSpace from ANSYS, Inc. and Pro/MECHANICA from Parametric Technologies. Today, design engineers can validate decisions within their primary design environment and make changes on the fly with little or no translation errors, or even translation steps for that matter. This is a significant advance over the days when features of moderate complexity ground the analysis process to a halt due to geometry import errors at the pre-processor level. The current high level of integration success has led many beginning users to believe that very little thought is required in the construction of CAD solid models for FEA. Since the most complex CAD models can now make the jump seamlessly into an FEA environment and mesh with very little effort, why worry about apparently non-value added steps like preparing analysis-specific geometry configurations?

The reasons for planning the CAD model to improve FEA throughput have changed over the years. Whereas 10 years ago, a design engineer had to consider what features would make it into the pre-processor and mesh, today, design engineers must consider what geometric construction best facilitates optimization, innovation, and efficiency. With schedules compressed to their practical limit, the decision to use FEA to validate a design versus simply going with gut feel typically hinges on the user's confidence that data can be calculated and interpreted in a timely manner. In many cases, the initial CAD model construction has the biggest impact on this factor.

In this chapter, we will review guidelines for maintaining a common model for both analysis and design and discuss CAD modelling techniques that can help or harm a user's ability to validate a design more effectively.

## 7.1 Linking Design and FEA Geometry

An early mantra of CAD implementation was that the software should enhance the design processes in place versus forcing companies to adopt new ways of working that were more convenient for the software. While the importance of this is self-evident, one must remember that with new tools come new opportunities. Adaptation of processes to take advantage of these new opportunities is an important characteristic of agile and forward thinking companies. The transition from 2D CAD to 3D solid modelling is a perfect example of this. To illustrate the point at a different level, consider the current popularity of screw guns or power drills outfitted with a #2 Phillips bit to allow construction with screws vs. nails. Hammers haven't gone away and nail-fastened wood structures are still a viable alternative. If a construction or remodeling company resisted the changes afforded by screws and focused solely on hammering technology, they would likely find their costs and schedules to be less competitive compared to companies that embraced changes allowed by the new technology. In this same manner, the increasing capabilities in today's CAD-embedded FEA tools are providing design engineering organizations in many industries the opportunity to adopt new processes that can improve their competitiveness. First and foremost is the ability to validate design decisions incrementally as a product takes shape versus waiting until the product has been completely designed. This requires CAD steps and techniques to be planned for validation.

First and foremost, a user must have a clear goal for the product behaviour and subsequent simulation tasks required to validate that behaviour. As stated previously, an important aspect of this goal is to understand the appropriate allowables that will indicate acceptable behaviour. More pertinent to CAD modelling is an understanding also of the part configuration best suited to provide that information. In coming to this understanding, consider the following.

### 7.1.1 Analysis Efficiency

It is a common misconception, most likely with its origins in the early days of FE modelling, that all small fillets or features should be removed from a CAD model before attempting to mesh. In a stress analysis, removal of some fillets or features can cause local results where that feature was to be unrealistically high. In stiffness or displacement based studies, some small features might have an impact on the results of interest as well. Similar to the process issue raised previously, a user shouldn't be forced to compromise the integrity of his or her model simply to satisfy the limitations of his or her FEA tool. However, the decision to include a small feature in the analysis should be conscious. The user should make sure there is enough mesh to capture the true nature of that feature and that the solution local to that feature has converged sufficiently. Failure to account for both of these disciplines will, at best, render the complexity required by that feature moot, or at worst, suggest to the user that the results are other than what they could have been.

That said, a user should consider the benefits of features that might be considered 'small' and decide if inclusion will pay off. If a feature is not included in the model, the user should be aware of the impact this missing feature might have had on the results of interest. This impact should be considered when the results are interpreted and the user should reaffirm the validity of this assumption or consider re-running the solution with that feature included to improve the local results.

Note that in discussing features of questionable benefit, the word 'inclusion' was used instead of 'suppression'. Another common misconception in CAD modelling for FEA is that insignificant features should be deleted or suppressed prior to analysis. Certainly, if they exist, this should be the case. However, there is little point in including them in the first place before it has been determined that the base geometry is sufficient. It is a natural tendency for design engineers to 'outrun their headlights' or to create more detail in their model than necessary before validating that their initial concept meets their needs. Adding design detail indiscriminately before confirming that a design direction is appropriate could result in wasted and/or duplicated work if the base features have to be re-considered. Thus, a more efficient approach is to abstain from including the feature in the first place.

All this discussion is based on the fact that a smaller mesh, in number of nodes, not necessarily element size, will solve faster. A model that solves faster provides answers to questions more quickly, thus allowing a user to ask more questions in any given amount of time. However, with the increasing speed of FE solvers these days, it is legitimate for a user to ask if evaluating, and in some cases defending, the appropriateness of a choice to leave out a feature might be more time-consuming than to simply include it in the model and taking the hit in mesh size. Resolving this contradiction will come more easily as experience with tools and interpreting analysis results grows.

### 7.1.2 Design Efficiency

As stated above, many users tend to build much more CAD geometry than is necessary before checking their initial decisions. This was referred to as 'outrunning your headlights', an allusion to driving too fast at night to react in time to something that appears at the limit of the headlights. A study of the features that truly control the acceptability of a design, as determined by the goals set forth previously, should provide a user some guidance to the minimal CAD model required to make sure a design concept is on track. This is sometimes referred to as the *Foundation Design*. It has been suggested that 80% of a product's cost and functionality is determined by decisions made in the first 20% of the design process. Should the Foundation Design be structurally flawed, any design work beyond the base features could be wasted time when the basic concept is revisited. An efficient design process is one where mistakes are caught early and wasted time and effort is avoided. FEA-based concept validation is an important tool in increasing design efficiency as every designer is empowered to check the impact of each decision on resulting performance.

Another way to look at the foundation design is as a sizing model. Since so much hinges on getting the initial concept correct and optimal, a designer can launch several quick studies on initial sizing of wall thickness, rib placement and cross-section, fastener quantity and placement, or other driving features. Integrated optimization tools can make the greatest impact at this stage. Once these base features are validated, it is reasonable to expect that short of a change to product definition, they should not need to be re-visited and subsequent design will not be in jeopardy.

### 7.1.3 Appropriate Idealization

The base features of the CAD model should correspond as closely as possible to the geometry requirements of the finite element idealization that will be most accurate and efficient for the problem at hand. Choosing the appropriate idealization was discussed in more depth in Chapter 5. If the part or assembly is predominantly comprised of long and slender components, beam elements may be most appropriate. If the parts in question are large thin walled components as is often the case in sheet metal design, shell elements may be best. If a 2D planar idealization such as plane stress/strain or axisymmetric will provide the fastest route to the required information, planning CAD to facilitate this should be considered.

#### 7.1.3.1 Beam Elements

If a beam element is deemed appropriate, somewhere along the line, wireframe geometry representing the placement of the slender members, either the neutral axis or a more convenient location that is corrected for in a neutral axis offset will be required. At the time of this writing, few packages have a solution for deriving beam elements from a CAD solid. Other codes will not be able to use solid geometry for developing a beam mesh so a user should start with a 3D sketch of the system before creating the mesh. Once the structural members and their cross-sections have been confirmed with simulation, they can be recreated in the CAD system using extruded features to represent each part, using the 3D sketch as reference to place the parts and drive their length.

#### 7.1.3.2 Shell Elements

The geometry of a shell element is most akin to a surface in CAD. As with beam meshes, if a shell mesh is the target for ongoing simulation, eventually, a surface representing the mid-surface of the thin walled component will need to be created or indicated based on faces of a pre-existing solid. Many FE codes now have automatic mid-surface extraction tools that can compress solid geometry into a surface body for meshing, either internal to the solver environment or explicitly in the geometry tools. Many codes also allow users to indicate an inside or outside surface of thin solid as the shell mesh location but this can lead to an inappropriate correlation between the analysis idealization and the intended geometry. Unless a user is also given the ability to instruct the solver to assume the midsurface is offset from the shell placement, the analyzed structure will be offset from the true shape of the part. These tools are not 100% reliable due to the wide variety of part shapes and solid modelling techniques. A more reliable method is to create the surface models that represent the idealization directly as the base feature to further CAD modelling. With this technique, a user can be confident that he or she can revert back to the analysis-specific configuration easily and reliably. In the plastic shovel example shown in Figure 7-1, a surface body is created in CAD. Validation and optimization is performed at this stage before any solid bodies are created. Once the overall sizing has been confirmed, the surfaces are 'thickened' into solids and additional design and manufacturing specific features are added to complete the assembly.

*Figure 7-1: Surface Model of Plastic Shovel for Shell Elements*

One side benefit of this technique is that it inherently resolves gap and continuity errors that often arise when compressing solid geometry to mid-surfaces. As shown in Figure 7-2, unless corrected for by some other means, there is a natural gap of ½ the wall thickness between the edge of one plate and the face of the other. Various software packages have different methods for resolving this discontinuity but they typically require user intervention. If a user doesn't notice this gap in the model, incorrect data could result.

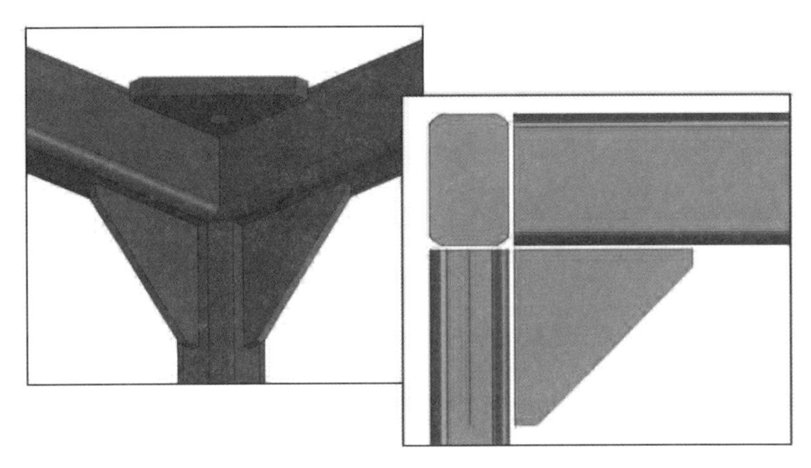

*Figure 7-2: Gaps Between Parts Compressed to Mid-Planes*

If your CAD system doesn't support conversion of surface bodies to solid bodies, it is important to model your thin-walled solids carefully with just the features required for the simulation up-front. You should check your compression periodically throughout the construction of the solid model for two reasons. First, it is easier to find problems when there are fewer opportunities for problems to occur. Second, with parametric CAD construction and parent-child complications, if one feature early in the model construction causes problems with compression and it happens to be the parent of many other features, resolving a mid-surfacing error could require much more work at the completion of the design than directly after that feature was created.

### 7.1.3.3 Planar Models

Most CAD geometry driven FEA codes require a planar surface or solid face for use as a *plane stress/strain* or *axisymmetric* reference. Many also require that this surface or face be positioned correctly, usually in the global XY plane. For axisymmetric models, the y-axis most commonly represents the centreline of the system, although some tools use the x-axis. A user should be aware of any such geometry restrictions required by their software prior to creating the CAD model so that interoperability errors can be eliminated.

Keeping a link between the design model and the analysis model for planar systems is the most difficult of the idealizations discussed herein. The fact of the matter is that a planar idealization typically represents only a portion of the system. However, on a case-by-case basis, a user should determine if a base sketch that can be optimized in FEA can also be used as an envelope or otherwise driving feature of the final solid geometry.

### 7.1.4 Boundary Conditions

A final factor in determining the configuration of the base CAD model to be used for both design and simulation is the boundary condition scheme to be used. Following the guidelines set forth in Chapter 6, a user should be able to determine if an applied load/restraint will result in the correct model behaviour or if additional assembly modelling is required to appropriately represent the response of the part(s) of interest. By going through the exercise of determining which parts are going to require validation and then what supporting parts are required to complete that validation before CAD modelling begins, the CAD work can be coordinated to ensure that all components interact appropriately. Sometimes, this will require that diameters of mating parts be made equal to facilitate contact. At other times, the placement of parts themselves can be adjusted to minimize meshing problems or simplify assembly interactions as shown below in Figure 7-3.

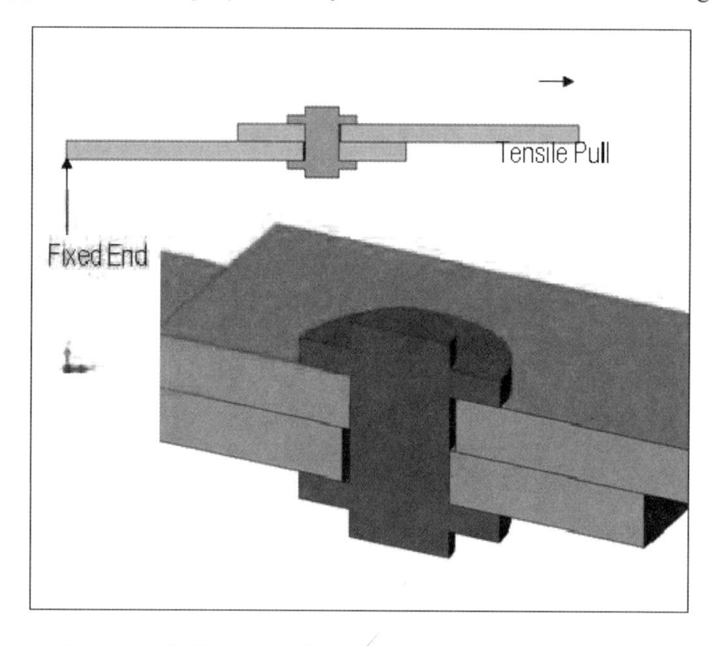


*Figure 7-3: Example of an Assembly Adjusted to Contact*

Keeping design and analysis models linked as a single database has obvious benefits. Many an analysis effort has been rendered junk by on-going design that invalidated the version being simulated simultaneously. Since the direction in the industry (and the focus of this book) is for designers to be performing some degree of analysis themselves, it is hoped that this wasteful scenario can be avoided. Another important benefit for keeping the models linked is optimization. Many CAD-embedded or CAD-integrated analysis tools allow for shape optimization studies that work directly with CAD features and dimensions. If a user optimizes on a model that isn't going to be used for on-going design, he or she will have to manually recreate the optimized configuration on the second model.

One downside of linking the design and analysis geometry is that the initial build time for the CAD model may take somewhat longer. In reality, the techniques required to construct a common CAD model shouldn't take more effort than other standard modelling practices. However, they may not be taught as rigorously or be as familiar to most users at first. Additionally, they require up front planning. Many companies have CAD modelling standards that dictate how best to proceed with a solid model for the purposes of downstream changes for manufacturing reasons. These may not support solid modelling for FEA and may even conflict with the techniques suggested herein. As with the hammer/screwdriver analogy made earlier, your company should review the processes in place for developing solid models in light of your goals for simulation and work towards a hybrid process that can take the best advantage of all tools available.

## 7.2 Impact of CAD Construction on Validation Efficiency

In the previous section, the impact of small features on solution efficiency was introduced. As stated, the speed of a solution is directly related to the number of nodes in a model. Small features can often force a noticeable increase in model size. However, one must ask the question, "How small is small?" Since everything related to size in FEA is relative, it is important to set a baseline for which to gage size on. The best measure for that is the nominal or default element size used by the software to mesh the majority of the part. Any feature an order of magnitude smaller than this size will force a local element size reduction and subsequent transition back to nominal. If this feature doesn't add any benefit to the quality of the calculated results, then the additional mesh it causes may not be worth the overhead. Additionally, features with small edges or that force the mesh size locally to deviate greatly from the nominal size can result in inefficient and possibly inaccurate meshes.

### 7.2.1 Meshing Basics

One possible outcome of including small features is that if the local mesh size isn't reduced correctly, the possibility of distorted elements arises. Understanding how most meshers work can help you understand why these factors are important. Most meshing software follows the same three steps. First, the edges on the surfaces of a part are evaluated and mapped with element spacing in whole numbers as close to the specified nominal element size as possible. When an edge is smaller than the nominal element size, one element is defined for that edge. Once the edges are mapped, or seeded, the surfaces are paved with temporary elements; triangles in the case of a solid part and triangles or rectangles (quadrilaterals) in the case of a surface/shell mesh. As described in Chapter 5, the ideal element shape for a triangular element or face is an equilateral triangle. Therefore, the mesher tries to create a surface mesh of equilateral triangles in all places where possible. When an edge that is much less than nominal length is encountered and no other meshing information has been defined the mesher will use that short edge as one side of an isosceles triangle with the other two edges being the nominal size. The further these triangles deviate from the ideal shape, the more error gets introduced into the model, at least locally. This error can be controlled using local mesh control but typically requires the user to see the small feature and manage the mesh manually. In some extreme cases, if the feature is much smaller than the nominal element size, the mesher can't complete the surface mesh at all and an error is returned to the user. Finally for solid models, once a clean surface mesh of triangles is completed, the volume is filled with tetrahedrons, again with the goal of having all 4 sides of the element equal at the nominal element size. When the thickness dimension between two surfaces is small compared to the nominal size, highly distorted tetrahedrons can be created which, in turn, introduce local error to the model.

In general, you should expect to have a good mesh on all the surfaces and features included in your solution. While an aggressive convergence process, either manually or automatically, should correct initially bad elements, it is in the user's best interest to review the externally visible mesh and note any areas of obvious distortion. Experience will be your guide as to whether you should correct the mesh before proceeding or simply monitor the results in those areas to determine if the data of interest might have been affected by this mesh.

From the standpoint of efficiency, any feature, planned or unplanned, that causes the number of nodes and elements to be greater than necessary can slow the solution down. For linear static solutions, this slow down may not be noticeable with today's software. For nonlinear and advanced dynamic studies, the delays can be significant. Good judgement coupled with experience can improve the overall process in time. A good rule of thumb however, is that if you can't explicitly justify the exclusion of a feature from a results standpoint, it is best to leave it in and pay the price in solution time. After the first few runs, you may be able to re-visit this feature for subsequent studies. However, do not jeopardize the validity of your solution needlessly.

## 7.3 Chapter Summary

Keeping the aforementioned in mind, there are some guidelines that all solid modelling practitioners can use to help ensure the CAD models created are as good as they can be for the purposes of efficient and accurate FEA. Since the focus of this book is for designers and design engineers who will be validating their own CAD models, at least initially, one important concept should be made clear up front. If you are analyzing a part that you created and run into meshing problems or lengthy solutions due to model complexity, you have only yourself to blame.

While it is in everyone's best interests to create a modelling standard that allows sharing of CAD models for all downstream purposes, you initially have responsibility and control over how smoothly and efficiently the simulation process goes and should take these guidelines to heart, if not for your team mates then for yourself.

# 8. Basic Solution Types & Their Limitations

An important characteristic of finite element analysis is that, in most cases, you only get the answer you asked for. In Chapter 3, you were introduced to the Key Assumptions in any finite element model that essentially defined the question posed to the solver: Geometry, Properties, Mesh, Boundary Conditions, and Physics. This chapter will discuss the physics aspect of your problem. You will typically indicate how the solver applies the appropriate physics by your choice of solution type.

For the most part, FEA solvers will assume you know what you are doing and will respond in kind. For example, if your product involves short-duration repetitive loading such that fatigue failures occur after several hundred thousand cycles, you should assess your model for durability. If you instead apply the load from a single cycle and ask your software to report static failure indicators, you won't get any push-back from the software. If buckling due to excessive compressive loads is the primary cause of failure, you won't be locked out of the linear static solver even though these results are unlikely to shed real light on the problem. Unfortunately, neither software nor hardware has reached a state where all possible physics can be examined routinely and simultaneously so users are required to decide which possible failure conditions might be applicable to their systems. Understanding what each is best suited for and what the limitations are can help guide you to the right choices.

## 8.1 Linear vs. Nonlinear

A linear static solution is the work-horse of the FEA world. It has been estimated that 90% of FEA is done with a linear static solution. In the world this book is targeted at, part-time design analysts, that number is likely higher. Interestingly enough, only a portion of the problems solved as linear static, possibly less than half, are actually representing systems and behaviors that are truly linear or static. Why has this technique become so prevalent? First of all, it's fast, cheap and easy. Linear static problems that once ground mainframe computers to their knees can now solve on laptops in minutes. Solvers that can handle these types of problems are very inexpensive and often offered free with certain CAD packages. None of that would matter if it wasn't for the second point. Many problems that aren't purely linear or static can be approximated as such with enough validity that design decisions based on this data are robust. In some cases, trend studies on problems with moderate nonlinearity can easily be made with the faster linear simplification. In others, the nonlinearity or dynamic aspect is neutralized by static testing designed to provide a gate for actual field response. In these cases, the analysis is simply modeling the test, not the actual operation of the system. So what does it mean to be linear and static and how can you determine if you've exceeded the limits of this approach?

Linear means that the resulting deformation (and thus strain & stress) is directly proportional to all the inputs such as applied loads, F, stiffness, K from material properties, geometry, and restraints. Thus, X = F/K such that if the load is doubled, the deformation is doubled. If stiffness is doubled, the deformation is halved.

### 8.1.1 Large Displacement

Linearity is only true for small displacement scenarios where increased stress doesn't alter the model's stiffness. A classic case of this is a guitar or piano string. As the string is pulled tighter, its resulting note increases, indicating a higher natural frequency caused by a stiffer system. The material or geometric properties weren't changed. Stiffness was increased solely by the increased string tension. Many structures exhibit this stress stiffening phenomenon, such as the paper tray shown in Figure 8-1 which is easily captured in a nonlinear large displacement analysis. However, a linear study will not redefine the model stiffness if this situation occurs because then the output would not be linearly proportional to the inputs.

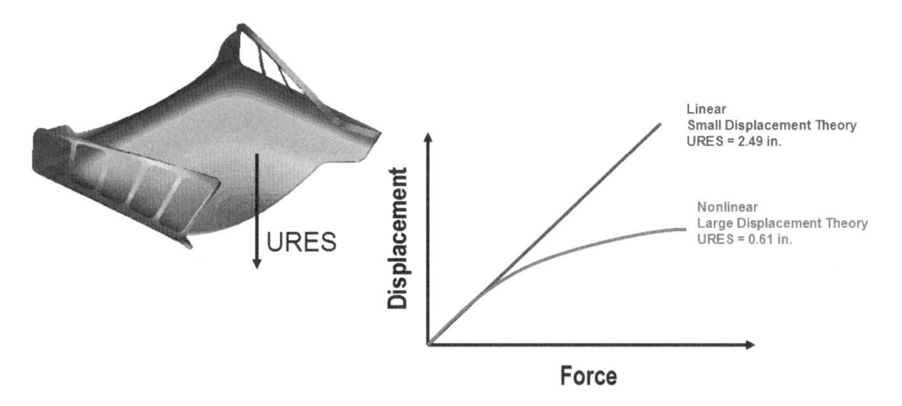


*Figure 8-1: Stress Stiffening of a Plastic Tray*

Another potential nonlinearity related to larger displacements is load redistribution or reorientation. For a linear analysis to be valid, it must be assumed that for the entire range of response, the load never changes magnitude, orientation, or distribution. A point load oriented vertically downward will always maintain this orientation and point of application, regardless of how much a part bends. In actuality, there is always a limit to how much a part can deform before you have to acknowledge the load has a new orientation or magnitude. For a linear analysis to be valid, the entire response of interest must occur before that turning point, or if some load re-orientation or distribution occurs, you must be convinced it will not impact the results of interest. If you can't confidently state this, a nonlinear large displacement analysis may be required.

### 8.1.2 Material Nonlinearity

The most common understanding of nonlinearity as it applies to mechanical structures is material nonlinearity. A linear analysis assumes that the entire response of interest occurs in the linear, or proportional, portion of a material's stress-strain curve. Many materials, such as most steels, have a reasonably large and well-defined linear portion of the stress strain curve such that a linear analysis for a system with stresses below the published yield strength is valid. Another class of materials have a small degree of nonlinearity to the elastic portion of the stress-strain curve that can be ignored as the impact this has on the elastic results is small or on par with the uncertainty introduced by a designer's knowledge of loads, geometry, or actual failure characteristics. A well-planned test correlation program can often wash out concerns about moderate nonlinearity. The stress calculated by an FEA model at a known failure load can be equated with failure even if the stress value doesn't match known allowables. If subsequent analyses are performed on similar systems, the linear stress should predict observed failure reliably, even though that stress may not be exactly what might be measured in the test part, should that measurement be possible.

A linear material model ceases to be sufficient when the nonlinear elastic portion of the stress-strain curve is pronounced. In these cases, a prediction on likely response needs to utilize the most appropriate material model, usually involving a stress-strain curve. While a trend analysis can tolerate substantial variation between predicted results and actual test results, a definitive statement on results based on a proposed design doesn't have that luxury. Another case is where post-yield response is an important aspect of the modeling. Whether you are being asked to determine the response after yielding has occurred or to calculate the spring back when a part that has yielded is unloaded, a nonlinear material model will be required. For either of these cases, it is best to consult with someone who has proven experience in these types of models before undertaking your first nonlinear material study.

### 8.1.3 Contact

Contact modelling is the indication of parts or surfaces representing discontinuous portions of the model that can interact with each other. Including contact has become more the rule than the exception in FEA due to improvements in both solving algorithms and computing speed. The need for more realistic part-part interactions were discussed in detail in Chapter 6. In its most basic form, a contact interaction in FEA is a conditional stiff spring between the nodes and elements on one part and those on another. As the two parts approach each other, the springs turn on, preventing penetration. As the parts pull away, the spring turns off such that it only supports compression. Hence, this is a nonlinear problem since the uniform proportionality cannot account for a spring that might be on or might be off.

As suggested in Chapter 6, some software packages have managed to implement a faster contact algorithm based on an assumption of small displacements. In addition to the restrictions attributed to small displacements in the previous paragraphs, another characteristic of a small displacement problem is that the change in orientation of one part to another as equilibrium is reached is negligible. Thus, the solver can save some compute time by examining the potential angle of engagement of potentially contacting faces and making some assumptions as to how they will resolve themselves as the load develops. Finally, a small displacement algorithm expects that when contact is established, the touching parts remain touching so that the solver doesn't need to re-examine the status of contact pairs after the initial evaluation. Essentially this results in a situation where parts whose contacting faces are all normal to the direction of applied load will solve accurately in a small displacement solution contact. As the contact normals deviate from the direction of applied load, error gets introduced into the contact calculation. Two such cases are shown in Figure 8-2.

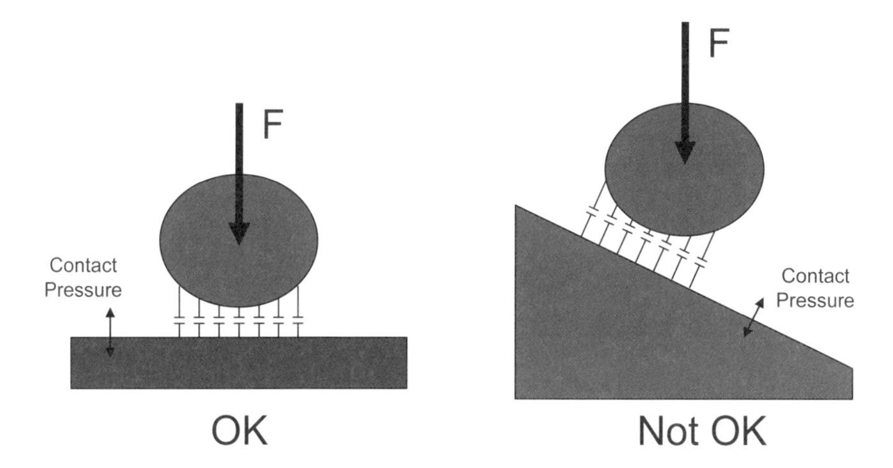


*Figure 8-2: Small Displacement Contact Limitations*

If your problem violates these conditions of small displacement contact, an invalid response will be calculated using these solving methods. The second case shown in Figure 8-2 is still a valid contact condition and will solve in most designer level FEA tools. It just requires a nonlinear large displacement formulation. A more dramatic case is shown in Figure 8-3.

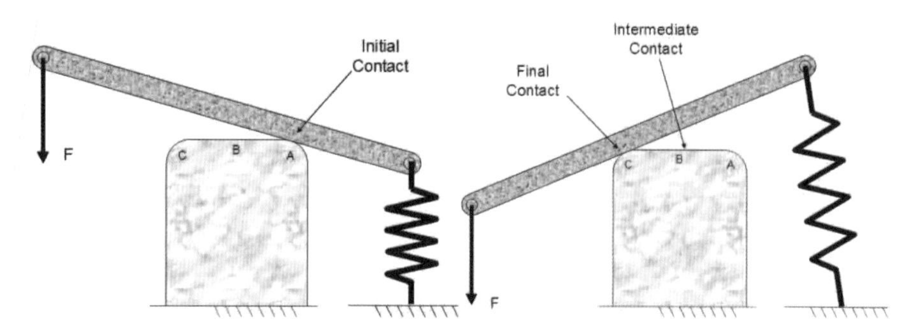


*Figure 8-3: Large Displacement Contact Example*

In the initial condition, the system has contact between the rod and block at Point A. However, as the applied load is applied the rod rocks over the top of the block (Point B) and reaches equilibrium with contact at Point C. In a large displacement contact formulation, the solver would solve the problem in load increments and assess the contact conditions at each step so that contact 'springs' that no longer need to be engaged are, in fact, turned off. Additionally, the large displacement algorithm makes no assumptions about what might or might not touch so the behaviour is allowed to progress naturally. In a small displacement solution, you might see the contact engage at Point A yet never truly disengage. You might also see a displaced shape indicating a correct response at the final position although the contact normal calculations, critical to the final stress solution, may reflect initial erroneous assumptions about where contact was supposed to occur. While it may look right, it probably isn't.

Simply put, small displacement contact can provide excellent results quickly if your problem corresponds to the restrictions measured and illustrated in Figure 8-2. However, not all codes make it clear when this algorithm is being used so if there is any doubt in your mind about the applicability, make sure you check at least one iteration with large displacement contact.

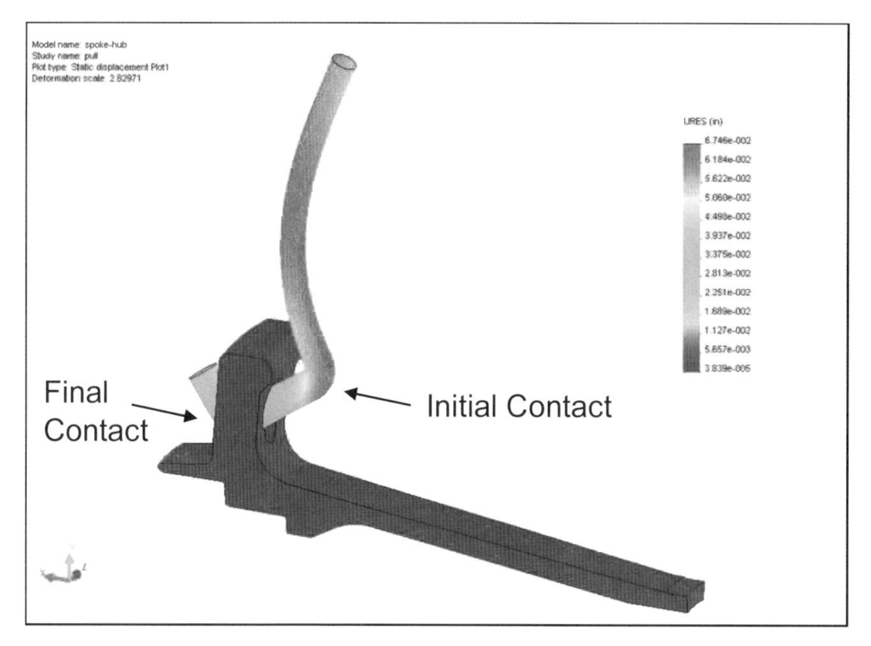


*Figure 8-4: Large Displacement Contact in a Bicycle Wheel Hub*

In the example shown in Figure 8-4 of a bicycle wheel Spoke interacting with a slice of the Hub (taking advantage of symmetry), a contact condition similar to the previous theoretical case occurred. The results on the Hub, just above the Spoke interaction, show (Table 8-1) how different the small vs. large displacement results can be.

| | **Max. Displacement** | **Max. Von Mises Stress** |
|---|---|---|
| **Large Displacement** | **0.0042 in.** | **53,000 psi.** |
| **Small Displacement** | **0.0039 in.** | **43,000 psi.** |

*Table 8-1 Results on the Wheel Hub*

To sum up the differences between linear and nonlinear analysis, consider using a nonlinear solution in the following cases:

- Your response involves material behaviour approaching or exceeding the yield strength of your material and absolute data at that point is important;

- You must make a definitive statement about a part's response when the stress-strain curve for the material indicates nonlinear elastic response in the stress ranges of interest;
- The deformed shape, viewed at true, or 1:1 scale, is significantly different than the initial shape such that loads may have reoriented or stress stiffening might be a factor. This is especially true for shell model idealizations;
- Contact interactions require non-trivial sliding and repositioning of parts from their initial placement to reach equilibrium.

If you haven't had much experience with nonlinear solutions, don't panic. Develop a resource to work with on your first attempts and be even more sceptical of the results than you might be with linear studies. For large displacement nonlinearities, the results interpretation of the final data is of similar complexity to a small displacement study. It just requires more computation time. For a material nonlinear problem, the material mechanics and failure characteristics get more complex so having resources to consult with, as you are learning about your material's post-linear behaviour, is highly recommended. NAFEMS "*Introduction to Nonlinear Finite Element Analysis*" is an excellent reference for this.

## 8.2 Dynamic vs. Static

As many nonlinear events can be approximated as linear, many time dependent events can be approximated as static. All events have some time element and your job as the design analyst is to understand how these impact the results of interest and the decisions you will make from those results. For example, all materials exhibit time-dependent material property variation. If a load is applied slowly, a part may deform through the elastic range and into the plastic range of stresses, possibly terminating in an equilibrium that renders the system inoperable yet intact. However, if that same load magnitude is applied rapidly, certain materials respond in a brittle manner resulting in fracture after only minor deformation. So the important question is, for the system of interest, how rapid is "too rapid"? Additionally, for many systems, static conditions simply don't apply at all. Machines with oscillating or rotating components, drop or impact tests, and vehicles subject to road vibration are all examples of systems where any question of time-dependency are moot. The dynamic nature of the system must be accounted for to properly understand the problem.

### 8.2.1 Static vs. Time-Dependent Systems

Static systems satisfy two key conditions. First of all, the load is considered to be applied slowly such that deformations are in a state of quasi-equilibrium for the duration of the application: inertial effects are non-existent. A simple example of this is the fishing rod shown in Figure 8-5.

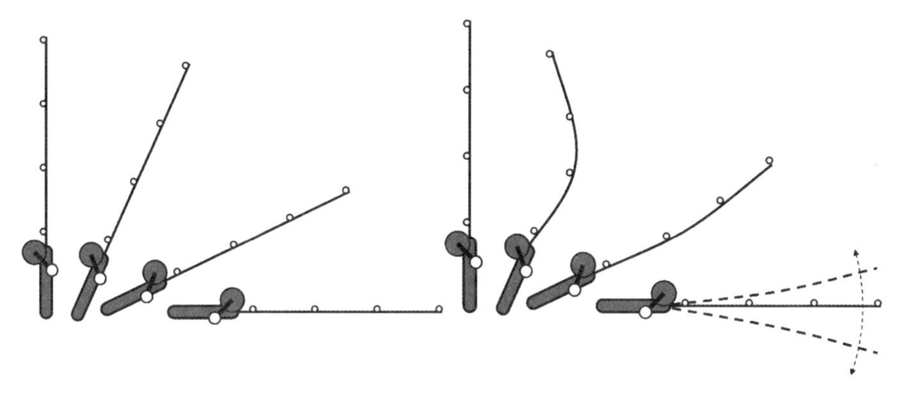

*Figure 8-5: Transient Response of a Fishing Rod*

The first sequence represents the rod being lowered slowly, possibly in hope of an easy catch right off the dock. The rod remains straight throughout the duration of the movement. The second sequence represents a casting motion where the rod is rapidly rotated (and the resulting whipping action flings the lure or hapless worm into more promising waters). The inertial effects on the rod cause a significantly different structural event. Virtually no deformation, and thus stress, exists in the slowly lowered rod. However, one can easily imagine how an older (or poorly designed due to misplaced static assumptions) rod might fracture if rotated too rapidly. Additionally, the rapidly rotated rod would most likely experience oscillations over a period of time until the various damping interactions brought the system to the same equilibrium the slowly rotated rod might achieve.

The second condition of a static assumption is that once applied, the load remains applied and never varies in magnitude. When the duration of the loading can affect the response of interest, a designer needs to consider the possibility that a transient (time-dependent) dynamic analysis might be required. A drop test is a common problem where load (impact load) duration can have produce a different response versus a statically applied load. As has been said previously in this text, your FEA tool typically assumes you know what you are doing and is unlikely to suggest time-dependent loading if you've otherwise indicated a load is applied statically, either intentionally or by default since no time-dependence was indicated. Fortunately, much like a nonlinear study, if a transient load is applied correctly and the response is comparable to that of a static study with the same load magnitude, then the static assumption is likely reasonable for further design work. Performing a nonlinear and/or transient analysis essentially hedges your bets with the fewest number of assumptions since most events are nonlinear and time-dependent. However, when these effects are negligible, a linear and/or static equivalent study is appropriate.

Product, test, and analytical experience can all help guide you in making the decision to pursue a static versus transient study. Many users of design analysis software may not have access to a transient solver. However, that doesn't relieve them of the responsibility to ensure that the static assumption is valid. If there are any doubts about this assumption, discuss it with others who are in a position to help discern the nature of the problem. A transient response can be less severe than a statically applied load as well as more so. Therefore, it is important to understand the system in question.

### 8.2.2 Static vs. Vibrating Systems

In contrast to the previous discussion on transient responses where equilibrium over a given time period is not assumed, many systems vibrate or oscillate at a constant frequency and amplitude such that they attain a predictable state of equilibrium. Hopefully, few designers would mistake these systems for static. However, instead of applying a sinusoidally varying load for a sufficiently long enough time period to achieve this equilibrium (although possible and occasionally desirable) these problems are best solved using a forced response or frequency response analysis. In these studies, a load is designated as oscillating and the solver calculates the peak response across one cycle of the oscillation. Some systems will calculate and provide stress and displacement data for fixed intervals within a given oscillation, typically reported as Phase Angle results.

Frequency response analyses are often performed over a range of input frequencies, or a frequency sweep. When the peak responses, stress or displacement, at selected locations in the model are plotted with respect to the input frequencies, peaks or spikes in the data will indicate critical natural frequencies of the system, in the direction of excitation. If a system is excited at a natural frequency in the theoretical case of no damping, the response is amplified infinitely such that collapse is imminent. Many factors contribute to damping such as friction, noise, micro-level material behaviour, or environmental effects such as vibration in a fluid. Thus all systems have some damping and the more damping present, the smaller the amplification, or attenuation, of the input signal at a natural frequency will be. The plot in Figure 8-6 shows a series of plots at different damping (indicated by $\zeta$ in the plot) levels. The x-axis reflects the ratio of input frequency over natural frequency such that maximum attenuation can be expected at the point where $X = 1$.

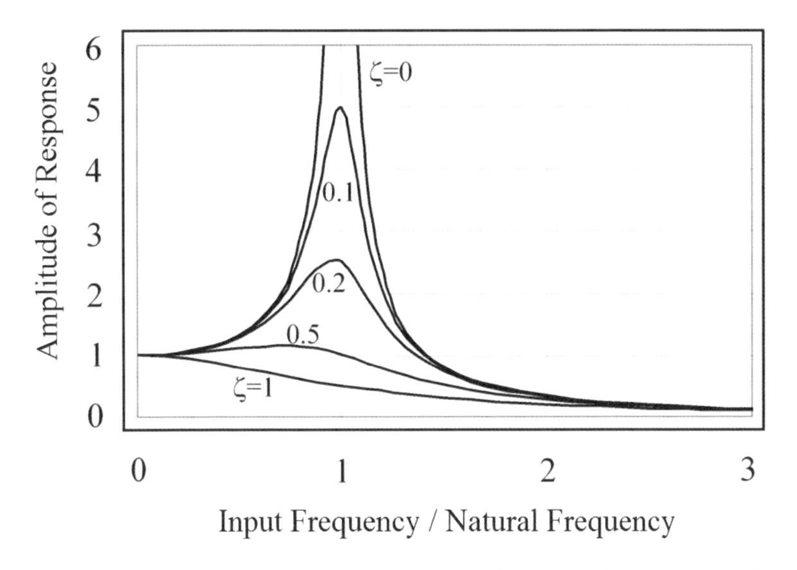


*Figure 8-6: Typical Dynamic Response with Various Damping Levels*

It should then be clear that when the system is operated at relevant natural frequencies, the likelihood for failure is increased. The frequency response study does provide stress and displacement values that can be used to make design decisions regarding acceptability (unlike modal analyses where the magnitudes are merely normalized to indicate regions of maximum and minimum potential displacement.) However, it is recommended that a modal analysis precede any dynamic investigation, either transient or frequency response.

### 8.2.3 Modal

Modal, or Frequency, studies report the natural frequencies of a given system and the shapes it would vibrate in if excited at each of these frequencies, or the mode shapes. A modal study doesn't require any loads and cannot, in general, take loads into account. Since vibratory deformation in a system is amplified if excited at a natural frequency, these studies suggest frequencies you should avoid in specifying motors or other dynamic inputs. However, in many cases, a designer has less control of the input frequencies than of the natural frequencies of the system so redesigning a system to avoid harmful natural frequencies is common. This technique is referred to as modal avoidance. To shift the natural frequencies in a structure, it is important to know a little about the factors that impact them.

Natural frequencies, typically reported in Hertz (Hz.), are the frequencies, with corresponding mode shapes, that a structure would naturally vibrate at. Recall the oscillation discussed in the fishing rod example of Figure 8-5. If you deflect the end of the rod and let go, it will oscillate at the 1$^{st}$ natural frequency. Natural frequencies are a function of system stiffness (including restraints, geometry, and material properties), weight, and the distribution of weight. This last factor is important. Figure 8-7 shows two systems with identical weights, stiffnesses and gross spatial characteristics.

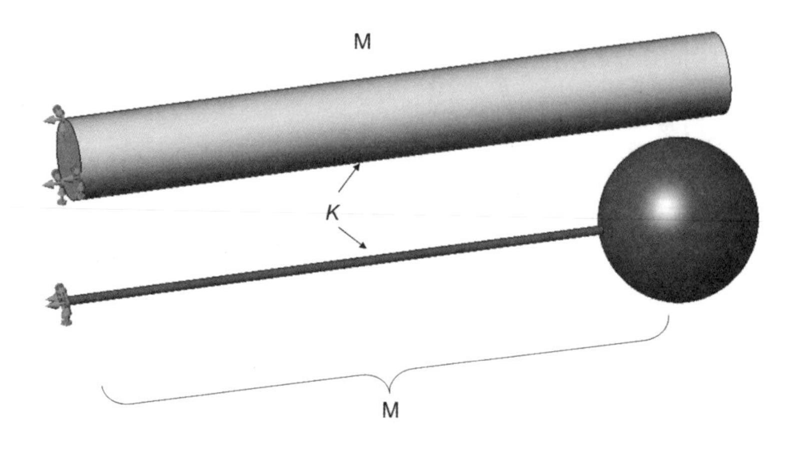


*Figure 8-7: Mass Distribution Example*

Dynamically, these are very different systems. The system with the distributed weight will have a higher first natural frequency since the center of gravity (CG) is farther from the fixed point. In the graph of Figure 8-8, you can see the first natural frequency of the system drops as the distance of the CG from the fixed end increases.

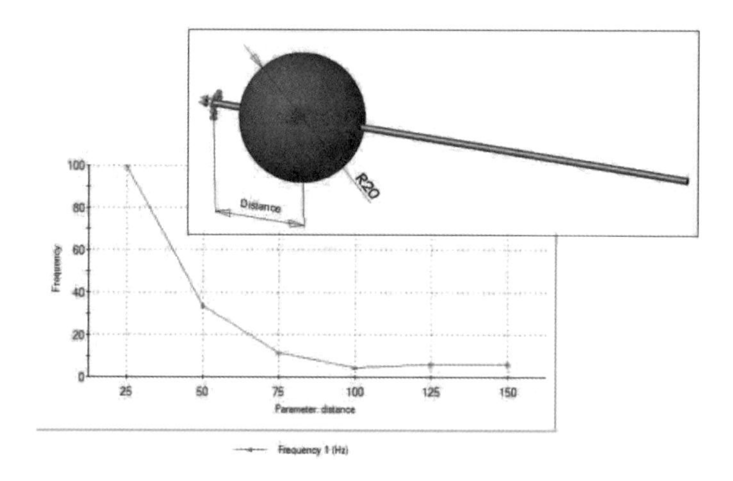


*Figure 8-8: Variation in Natural Frequency with Mass Distribution*

In general, the natural frequency can be considered a function of K/M, or stiffness over mass. As the stiffness increases, the first natural frequency increases which is typically desirable. However, as the mass increases, the natural frequency decreases which is less desirable. Unfortunately, the most common techniques for increasing stiffness in a structure typically involve adding mass so the impact of such a change is potentially null. Recalling that natural frequency is affected as much by distribution of mass as the mass or stiffness itself, you can choose to stiffen a structure with components that add mass closer to the fixed locations than farther such as in the example shown in Figure 8-9 where a small amount of mass, M, correctly placed, far exceeded the dynamic stiffening offered by the brute force method of simply increasing wall thicknesses. Note that the dimensions in the images denote the wall thickness of the vertical members, not the relative displacement. A modal analysis does not include the magnitude of the input excitation so it cannot respond with a response magnitude.

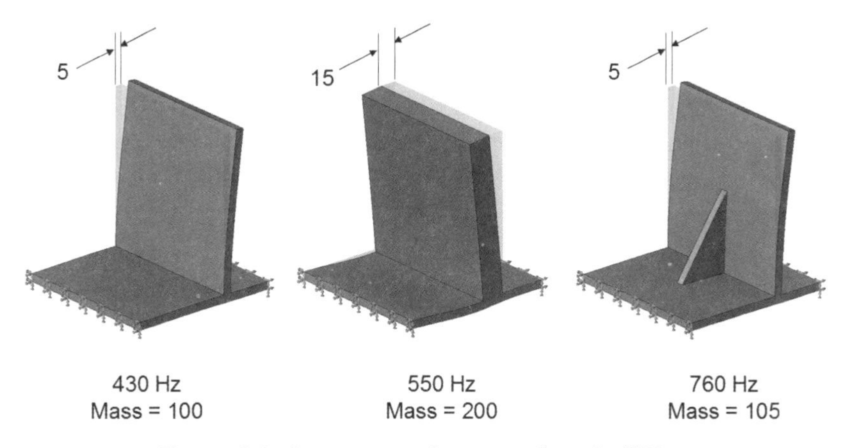


*Figure 8-9: Optimization Iterations for a Stiff Plate*

Remember for a modal study, any displacement or stress magnitudes reported by your system are only relative values and should not be mistaken for actual amplitudes your system might see. To understand the actual response of your system if shaken at a given frequency, you'll need to employ a frequency response dynamic study as described previously. These are more advance studies, both analytically and from an engineering standpoint since the results of interest are extremely sensitive to damping and component interactions. You should plan on working with someone who has proven experience in these problems before attempting to take one on yourself.

## 8.3 Drop Test

While technically a dynamic analysis, drop test studies, or "drop on" studies (where a weight or projectile is dropped on a product) have been growing in importance in product design, especially consumer products, since this test may drive material, fastener, or wall thickness decisions far more rigorously than any operating condition.

Drop testing is a costly part of product development for a couple of reasons. First, it is a destructive test and, if done correctly, involves destroying several units to ensure a safe statistical sampling. Second, since prototypes can't always be expected to respond to drops as finished, tooled, injection molded parts, it typically happens at the very end of the product development process when everyone is hoping to be done. When a problem shows up at this stage, it initiates a high-visibility fire drill to try to solve the problem as quickly and as cost-effectively as possible. Acting quickly and with minimized product cost impact usually must be offset with high development costs in tool modifications and engineering time. This is a scenario where virtual testing can really pay off.

It is worth noting that detailed drop test studies of an entire product typically involve the most complex nonlinear and dynamic techniques. Specialized software that can account for short duration, strain-rate dependent, nonlinear material properties, and complex contact are required. Preloaded fasteners & components may also need to be considered. Part-time design analysts should not embark on a drop test of a multi-part assembly lightly. You are highly encouraged to work with an internal or external consultant on your first few projects to debug your material properties and assembly interactions. You should rigorously correlate analytical results to test results and adjust your processes based on this correlation. Once you've got a few of these under your belt, you may feel comfortable enough to go it alone.

Where design analysts without the benefit of a mentor can take advantage of drop test techniques is (again) in a trend study mode where potential improvements are evaluated against a baseline. This works best for single parts, continuous load bearing assemblies, or assemblies with only a few parts that have very predictable interactions. As in static trend studies, the actual magnitude of the responses should be looked at only in relation to the magnitudes of other iterations.

Some other important considerations when planning drop or drop-on simulation is that static equivalent studies for these highly dynamic events are very difficult to get right in a general sense. A common static "equivalent" technique is to fix a part on a face furthest from the point of impact and apply an equivalent static force to the part at the impacting surface. This is sure to provide misleading results since it negates the inertial effects of the impact on the rest of the system. A closer yet often equally misleading technique is to restrain the point of impact and apply a gravitational body load equivalent to the deceleration the part would see on impact.

This last technique has the best chance of succeeding if the center of the impact force vector passes through the CG of the part or assembly and you are, again, working in Trend Mode. Due to the dynamic nature of the event, a static load (recall the assumptions inherent in a "static load" described previously) may over-predict or under-predict the actual response so trends are the only valid output from this study. Additionally, if the CG is not in line with the impact, product rotation will occur and there could be secondary impacts which might be more damaging than the initial study. This is illustrated in Figure 8-10.

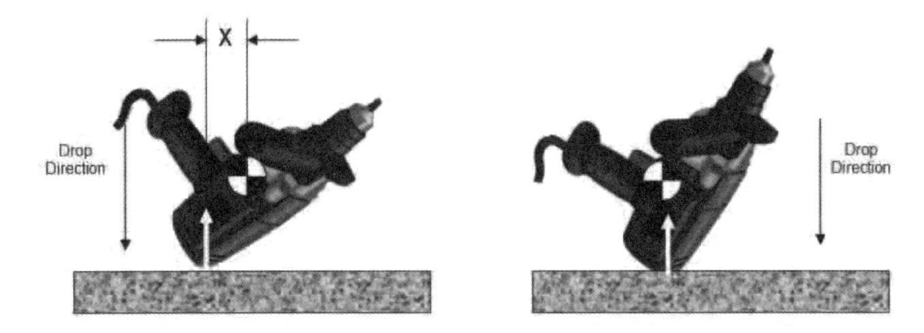


*Figure 8-10: Example of CG Alignment in a Drop Test*

With an equal distribution of mass on each side of the impact location, as shown in the right-most image, the system is quasi-stable. However, in the left-most image, it is unlikely the product will remain in the impacting orientation long enough for any substantial load to develop as impacted. If you think you might still be able to take advantage of this technique, remember that the impacting force is related to the stiffness of the impacting parts. Stiffer parts have a greater impact force and vice-versa. If your design improvements affect the stiffness of the part local to the impacting features, you may need to re-calculate the impact force applied.

The most reliable approach is to model the impact event with a dynamic analysis. Some design analysis tools have pre-defined drop test applications where users can simply input drop heights and impact orientations before initiating a solution. Others require more complex inputs and contact definitions. Either way, make sure you start with some simple problems that have intuitive results & move to more realistic parts that you can correlate fracture locations with physical testing before attempting to draw conclusions on more complex systems.

## 8.4 Chapter Summary

While the majority of analyses performed by design engineers will be linear static, it is critical that all practitioners of FEA understand the limitations of this solution type and the solver needs of their products and tests. If your product behaviour, or more specifically, the data that allows you to qualify your product behaviour, requires nonlinear, dynamic, or buckling analyses, don't attempt to circumvent the process with the wrong tool. You would be better off relying on physical testing than to mis-apply analysis intentionally. Make sure that you can address your simulation needs with the proper solution capabilities.

# 9. Simulation Model Verification

At this point in the text, readers should be aware that there are many ways to construct a simulation model for any given scenario. Some will produce relevant and meaningful data and others will produce correct-looking yet incorrect results that can lead to erroneous design decisions, increased product cost, and/or more prototype iterations than might have been saved had the analysis been performed correctly. Following the guidelines put forth in the preceding chapters should empower you to make better decisions on modeling options and assumptions and to understand the limits of the data provided by your models. In this chapter, some additional checks will be offered to help make sure nothing slipped through the cracks both before kicking off a solution and after the solution has completed. Many of these are techniques used regularly by expert analysts to check their own work and should be part of your day to day work in FEA.

## 9.1 Pre-Analysis Checks

Performing model checks before starting an analysis is important for a couple of reasons. First of all, many analyses take more than a few minutes to complete so if you can make sure more lengthy solutions are valid beforehand, time can be saved in the long run. Secondly, there is a common tendency to believe what's on the screen once the analysis has completed. Graphics in today's design analysis software have become so good that the data presented is very compelling. If a modeling mistake makes it through the solver, users often don't look back. Finally, once feedback on model performance has been offered, it is likely you'll start looking for improvements right away which means making changes and solving the model again. Any mistakes in the set up of the model will propagate through all successive iterations. For these reasons, it is worth taking the time to walk through some of these simple checks.

### 9.1.1 Inspect the Mesh

One of the easiest checks you can make is a visual inspection of the mesh. Turn off all load, restraint and mesh control symbols and any other entities that obstruct your view of the actual mesh on the part or assembly. Look at the mesh shape on curved surfaces and the transitions between smaller and larger elements. If these don't look smooth and continuous, the results in these areas will be suspect. Make sure that the mesh captures the nature of the features. Round holes shouldn't look like squares or hexagons. Fillets shouldn't look like chamfers. Another good check, available in some tools, is to view the geometry and the mesh simultaneously but in different colors. If patches of geometry are visible across multiple continuous elements, the chordal error in that area may be too great to represent the actual part.

If your tool supports it, display different materials in a model with different colors. This will help you quickly identify any mis-assignment of material prior to running the study. Similarly, if you are using shell or beam elements, use the tools in your modeller, if available, to display all elements of similar properties in the same colors. For example, when all shells with 1mm thickness are displayed as blue and all with 2mm thickness are displayed as green, any mistakes in wall thickness assignment become obvious. This can be done with similar benefits for beam elements. Another tool only available in some software packages is to plot shell and solid elements as their 3D equivalent to get a better visualization of how the solver will interpret your mesh.

A final check available in most tools is to identify less than ideal shaped elements in your model. If your tool supports it, display a mesh quality plot to visually locate elements that might produce erroneous results. Element aspect ratio is the easiest and most common distortion type to check. If your tool doesn't have a visual display of mesh quality, you should still be able to list out the elements with poor quality and locate them using XYZ coordinates of nodes or other manual methods. If the worst elements show up in areas you don't expect to be a concern, you may just want to note these locations and continue with the solve. If these areas present a stress issue once the run is completed, you may want to remesh the area to clean up the problem. If you know that the distorted elements are already poised to add error to critical locations, fix them before solving.

Remember that the FEA solver only understands your geometry by the mesh representation so if the mesh doesn't look like the part, the analysis results may not represent the behaviour of the part. Even though automatic meshers are making mesh quality concerns less of an issue, user error on complex models can still result in mesh related problems so a quick check can't hurt.

### 9.1.2 BC Checks...Again

Another important check is your load and restraint application. Before kicking off the solution, review your graphical symbols visually. Are the arrows pointing the right way on loads? Do pressures on shells all point in consistent directions? Are the load and restraint symbols showing on the right entities? Also run a quick check of your input definitions for each. Are units defined consistently, or at least correctly? Are loads defined as "per entity" values sized correctly for the number of entities selected? Are any gross magnitude errors obvious, such as a decimal point mistakes?

If your tool supports a check of total applied load prior to the solution, verify that the sum of the forces applied to your model makes sense. This is an easy way to catch missed entities, units, or decimal places.

Check your restraint scheme to make sure there is a path to ground in all spatial degrees of freedom. Do a quick sanity check to see if you are accidentally over-constrained.

### 9.1.3 Material Properties

If you used material properties from your tool's library, do a quick check of the input and failure properties. If you defined a custom material, perform a more thorough check. Are materials assigned to the correct parts in a multi-part, multi-material model? Always make sure you check the units for material properties you assign, whether you pull them from a library or define them yourself. This is a very common mistake users make.

### 9.1.4 Model Mass

If your software supports it, check the mass or weight of the mesh and compare it to the mass or weight of the geometry. If you are only using solid automeshed elements, there should not be a noticeable difference. However, as you incorporate more idealizations and virtual structure in your model, the mass differences could become significant. This is a good way to make sure that all the parts that need to be meshed are meshed. Additionally, if you are running a dynamic or modal solution, or if you are using body loads like gravity or acceleration, the mass defines the load so you need to make sure this is right to get meaningful results.

Mass deviation is most likely to occur at intersections of shell or beam elements. Failure to account for the weight of weld beads can also lead to an under-prediction of mass. Failure to include or properly account for fasteners such as bolts, pins, springs, or hinges may also adversely impact the mass calculations on the mesh.

If the difference between the expected mass and the model mass concerns you, you can add mass by correcting entity definitions or add a layer of idealized elements to represent the mass itself. Most tools allow you to input lumped mass elements at prescribed points. While the intent of these is to represent purchased or non-essential yet heavy parts, they can be used for "make up" mass. Some tools even allow the inclusion of non-structural mass to certain element types. This property essentially makes a group of elements heavier than the volume & density would suggest for just this purpose. Avoid simply increasing density of materials in the model to make up mass unless you are confident this mass is uniformly distributed in all volumes.

### 9.1.5 Mesh Size vs. Time and Resource Constraints

One of the last checks to consider is model size vs. constraints on your schedule or resources. Some FEA tools provide a run time estimate at the beginning of the solution. If you typically deal with larger models and/or tight deadlines for your problems, it would be a good idea to note model size, predicted run time and actual run time in a journal along with the workstation the job was completed on. Include this data in the report for future reference. As you gain experience in model sizes and run times, you should be able to estimate whether your model, as constructed, will solve in the available time or with the available hardware restrictions. It's better to know before you start than to walk away from a long run expecting promised data and then having to scramble a few hours later because the solution ran out of disk space or was still going with no end in sight.

## 9.2 Post-Analysis Checks

No matter how diligent you were setting up the model, there are still things you can't check until the solution completes. Some of these will relate to assumptions you made but weren't entirely confident of (which should be the majority of them) and some will be related to the unknowns that comprise the goal of the problem itself.

### 9.2.1 Solver Report

Most solvers produce a summary report at the end of a run that provides a wealth of information to users who wish to delve into them. Noting the number of solved degrees of freedom compared to the disk & RAM used as well as the solution time can help in planning future analyses. Checking for any error messages that suggest, in the language of your solver, that a less than ideal solution was reached is important. Similarly, this report should indicate the success (or failure) of an h-adaptive solution if the user interface doesn't inform you directly. This report may also summarize the sum of applied loads, maximum or minimum response values, iterative error results among other things.

### 9.2.2 Sanity Check on Displacements

The next thing to look at when the solution completes is the displaced shape of the system. Is it bending where it should? If you expected displacements on the order of 1mm and it displaced 100mm, it is likely something didn't work as you thought it would. Examine the color contour displacement plots on assemblies to verify that displacements are continuous at bonded or fastened interfaces. Are parts that you intended to be attached displacing as if they are still attached?

Animate your displaced results. Watch the entire part deform in an exaggerated scale. For contact problems, remember that a linearly scaled exaggeration will show contacting surfaces penetrating since the math behind contact formulations does allow small amounts of penetration. Make sure you look at the entire model, in these cases, at a 1:1 scale. Examine the animated displacements at each load or restraint individually in an exaggerated view for all models to make sure they are behaving as the parts the boundary condition represent would have caused them to behave. This is the easiest way to spot poor load or restraint decisions.

Remember that all results in FEA are based on the strain solution which is best validated with the displacement results. If the displacements don't look right, there is a good chance your stress results will be misleading.

### 9.2.3 Review Contact Regions

At each contact region, examine penetration to ensure that settings in your solver didn't permit too loose a contact definition. This is best viewed at 1:1 scale and section plots can help zero in on the critical areas. Seasoned analysts have many tools at their disposal to query the validity of contact regions. However, I'd suggest that if the contact doesn't appear objectionable using the aforementioned visual examination, then penetration is probably not an issue. However, there are other ways contact algorithms can miss. If your code supports it, examine contact pressure or stress plots for each contact region. The distribution should be reasonable and consistent. Just as in your examination of stress contours to evaluate convergence, there shouldn't be "hot spots' of high contact pressure in the middle of regions of lower contact pressure unless the geometry indicates them. If contact force vectors are available, make sure these are all normal to the surface and pointing out of the part. As silly as it sounds, a common contact calculation error is to have a contact element within a contact region pull instead of push. A couple of inverted contact elements can cause error in the load path and resulting stress calculations.

If your software supports contact but not contact pressure plots, a P3, or Minimum Principal Stress plot can often provide similar insight. P3 is the most compressive principal, or normal, stress on a given node. If contact is performing correctly, this stress quantity should track contact pressure pretty well in most cases.

### 9.2.4 Reaction Forces at Restraints

Most FEA tools allow you to query the reaction forces at restraints. This is an important check, especially if you were diligent enough to complete a free-body diagram prior to running the problem. The reaction forces at restraints should compare well to the reaction forces calculated in the free-body diagram. If you didn't complete a FBD prior to building the model, a check of the reaction forces provides one more data point to make sure nothing out of the ordinary happened in the completion of the study.

### 9.2.5 Convergence and Error Estimates

Once you've validated that the general model stiffness and confirmed that the load path at both the restraints and contact regions is reasonable, check convergence.

This will ensure that the mesh is sufficient to capture the stresses of interest. Techniques for checking convergence were reviewed in detail in Chapter 5. This is the point in the analysis process where these checks become important. Use all of the techniques provided to make sure that the mesh on a single analysis has converged sufficiently to make your interpretation of stress results meaningful. You also need to make sure that the convergence levels on successive iterations are consistent so that comparisons are valid. Stress variability due to inconsistent convergence can easily be on the same magnitude as those due to geometric improvements.

## 9.3 Chapter Summary

Checking your model is one of the most important tasks in the whole analysis process yet too few design analysts take the time. Unfortunately, this leads to wasted effort, at best, or mis-leading design decisions. Getting in the habit of following these checks will pay off in the long run after only a few modelling mistakes are caught.

# 10. Closing the Loop: What Does it All Mean?

After all this hard work, it may be disconcerting to know that getting the answers and validating them is only half the battle. The most accurate data in the world is useless if you don't know what it means. This is where design and engineering experience needs to take over in the project. While the vast majority of this book has been focused on problem set up, the fact is that for most design analysis users, these first tasks will be completed in minutes once familiarity with the interface and basic assumption validation has been completed. The second phase of the project is where you are most on your own. Even within a given industry or across different companies developing directly competing products, the engineering need may vary greatly such that nearly every reader of this book will need to interpret their results differently. Consequently, what may be the most important part of the process gets a disproportionately small portion of the text since there is no conceivable way to capture all the diverse needs of the readership.

What we will try to do in this chapter is provide some guidelines on results interpretation and suggest additional explorations you might undertake to make the data more meaningful for your design needs. This will include a discussion of test correlation as well as an introduction to sensitivity analysis.

## 10.1 Choosing Meaningful Output Quantities and Displays

Based on your project goals, hopefully determined prior to starting your analysis, you must decide what output to review. This will be over and above the output suggested in Chapter 8 to validate your model.

### 10.1.1 Structural Failure

If you are primarily concerned with stiffness where acceptability is determined by the magnitude of deflection or deformation, the resultant displacement should be your output of interest. You may also want to consider looking at displacement along axes of global or even user-defined coordinate systems (CS) if these aid your understanding of the problem. The most common alternate CS displacements are the radial and tangential components of a cylindrical CS. If you have a component spinning at high speeds that has been press fit on a shaft, the radial displacement will tell you if you are in danger of losing your press. A tangential displacement plot will help you understand wind up in shafting or other torque bearing components.

If failure by yielding is a concern, then it is expected your material is ductile and you should examine a Von Mises Stress plot. Von Mises Stress is the most reliable predictor of the onset of yielding for steels and many other ductile materials. Von Mises Stress doesn't provide any indication of tension or compression in the model so it may be helpful to also review a Maximum or Minimum Principal Stress plot since these do provide insight to the tensile or compressive nature of the stressed area. If you recall, Maximum Principal Stress, or P1, is the most tensile normal stress at a given point in a model where Minimum Principal Stress, P3, is the most compressive normal stress at a given point.

If your material is more brittle in nature, then you should review output that indicates fracture since brittle materials don't yield appreciably before failure. P1 and P3 stress plots should be compared to the Ultimate Tensile Strength ($S_{UT}$) or Ultimate Compressive Strengths ($S_{UC}$) of your material, respectively, to gage the potential for fracture. If your software permits it, an even better indicator of brittle failure is the Coulomb-Mohr Criterion which is a quantity that combines P1, P3, $S_{UT}$ and $S_{UC}$ to provide an indicator of safety. If the Coulomb-Mohr criterion is less than or equal to 1, failure should be expected. If greater than 1, you should be OK, subject to any other safety factors you have deemed appropriate for your design.

Predicting fracture in ductile materials requires a nonlinear material analysis. A common mistake is to compare Von Mises Stress in a linear material analysis to Ultimate Strength for the purposes of predicting fracture. Keep the following guidelines in mind when evaluating your results.

A linear model of a ductile material predicts onset of yielding. Once the applied load exceeds that which causes yielding, neither the deformed shape or stress solution is valid. This is analogous to taking a part in test to the point where measurable yielding has occurred and then stopping the test. At this point, you can expect good correlation between your analysis and test results. If you take the test to failure without examining the parts at the onset of yielding, the best comparison you can make is to hope the most deformed areas correspond to the regions in your FEA model where stress is the highest. Due to stiffness changes and load path re-distribution, even this isn't guaranteed.

A nonlinear material model with plasticity can capture widespread yielding up to the onset of fracture. If you stop your test at the point where the first cracks become visible, you may be able to get good correlation to your analysis model. This is especially hard because many cracks initiate and propagate very quickly once a load threshold has been exceeded by even a small amount. Additionally, many cracks first appear on hard to examine features or at the contact interface between loaded parts. However, once cracks occur in the model, the stiffness and load path once again change and any additional applied load reduces the chance of correlation to your analysis model.

To capture post-fracture response in FEA, specialized software with specialized elements are required and you will have moved out of the realm of the design analyst and into the realm of the specialist. To get meaningful post-fracture results requires a skill level that few part-timers will acquire based on the hours of experience and research required.

### 10.1.2 Modal or Frequency Failure

If the response of interest is natural frequencies because you want to make sure your system doesn't respond objectionably to vibrational input or if you want to redesign your system so there are no natural frequencies near driving frequencies, you should examine the natural frequencies and resulting mode shapes from a modal analysis. List or print out all the frequencies within the range of interest. Note groupings of frequencies that vary by only a few percent. It is likely that an input frequency near this grouping could excite any or all of these. Examine the mode shapes of natural frequencies near known excitation frequencies. These are best viewed as animations. The default exaggeration scale should be sufficient since the magnitudes of modal analysis deformations are not meaningful as discussed in Section 7.2.3. Mode shapes that can't be excited by the direction of the applied excitation are not likely to be a concern as illustrated in Figure 10-1.

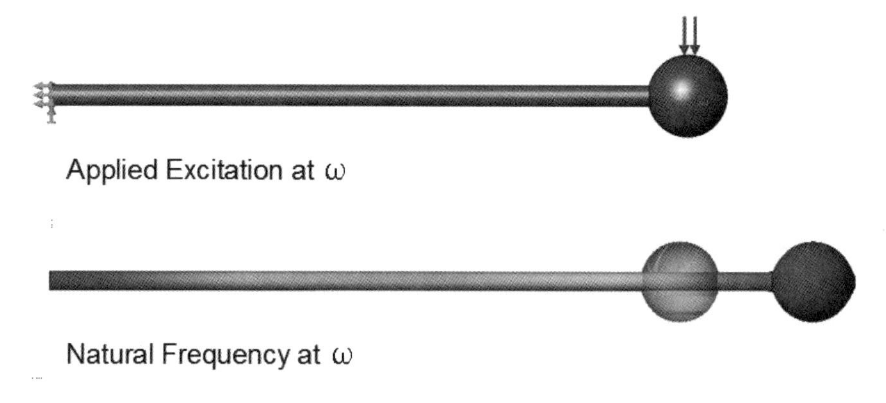


*Figure 10-1: Difference in Mode Shape vs. Excitation Direction*

The mode shape (axial extension) of the natural frequency that corresponds to the excitation load, (bending), is unlikely to be excited even if the system is driven at this exact frequency. When making design decisions involving modal results, make sure you look at both the frequencies and corresponding mode shapes.

The simple example above is an extreme example. You should also note that a mode shape calculated by a modal analysis represents excitation by a perfectly oriented load. In most cases, vibration input is "dirty" in that it contains multiple orientations, oscillations, and frequencies. Before you rule out a mode shape and corresponding natural frequency as irrelevant, make sure you know that it can't be excited by minor eccentricities in the load, alternate installation orientations, or miss-use.

### 10.1.3 Buckling Failure

Buckling can occur at loads and stress levels well below the failure strength of a material so stress is not a good indicator of potential buckling. If you look at P1 & P3 and see that the stress field is primarily tensile so that the risk associated with inadvertent buckling is small, you might be able to assume buckling won't occur. However, a buckling analysis, which provides a buckling load factor (BLF) or safety factor is the best tool for understanding the possibility of buckling failure. A buckling load factor is essentially a multiplier on the applied load to determine what the first or critical buckling load is. If you applied the maximum load expected, then this truly is a safety factor where a BLF at or less than one indicates buckling is likely. In reality, due to the non-conservative nature of buckling, a safer approach is to shoot for a BLF greater than 3 before ruling buckling out.

For physical as well as numeric reasons, a linear buckling analysis is non-conservative, hence the recommendation of a BLF of 3. Additionally, these studies will only indicate the onset of buckling and provide no data on what will happen after the buckling initiates. Some structures collapse completely while others self-stabilize. If you need more precision in your buckling results, consider running a nonlinear large displacement analysis. A nonlinear solution will capture buckling without any additional input, (unlike a linear static analysis which will never indicate buckling.) One good indication that buckling is occurring in your nonlinear model is when it diverges at the same applied load, regardless of the convergence settings you try. Solving through a buckling instability is extremely difficult for an analysis solver. Forcing a solution past the onset of buckling can be very time-consuming and often for very little potential gain. In most cases, the material, contact, and general chaotic nature of buckling means that any post-buckled results are still no better than a guess or one possible post-buckled outcome. However, a nonlinear analysis used in this fashion can provide a more accurate indication of the true buckling load and when, after subsequent design improvements, the solution completes to full load, you can be reasonably confident that the buckling problem has been corrected. If this type of study is crucial, consider using nonlinear material properties and experiment with eccentricities in assembly and load orientations. A perfectly square assembly is less likely to buckle under compressive loads than one that starts somewhat racked to the side.

## 10.2 Safety Factors and FEA Results

Most of the previous discussion, and, for that matter, most discussions of FEA results in general, discount safety factors. A common misconception is that FEA results are more accurate than previously obtained results, thus minimizing the need to worry about safety factors. It is important to remember that safety factors account for uncertainty and typically contain three primary components. The first is calculation accuracy, which a properly completed analysis with valid assumption and results interpretation can minimize, although not eliminate. The second is uncertainty in manufacture which FEA can only impact if manufacturing variations are explored in alternate FEA models. This isn't often pursued in design analysis tasks. The third is uncertainty in use. Analysis can also help you understand this better if predictable misuse is explored. The fact of the matter is that uncertainty of use or manufacture are not likely to be any better understood with access to analysis than they were prior to FEA and, in most cases, represent the majority of uncertainty so any and all safety factors should probably be applied to the analytical results as well.

## 10.3 Factoring in Sensitivity and Uncertainty

An important tool for understanding design variability and uncertainty is a *Sensitivity Study*. This is a structured study of a variable, looking at how much a result of interest changes based on changes to the input variable. Many tools provide automated ways to explore sensitivity including specification of the variable input range (min to max values), increments, and results plotting. The tool for exploring sensitivity in COSMOSWorks from SolidWorks Corporation is called Design Scenarios. The input form is shown in Figure 10-2 and output for the four independent dimensional variables defined in the study is shown in Figure 10-3.

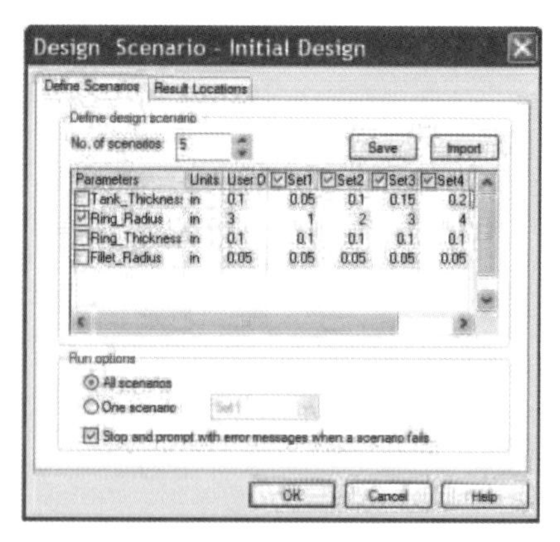


*Figure 10-2: Typical Design Scenario Set Up for a Sensitivity Study*

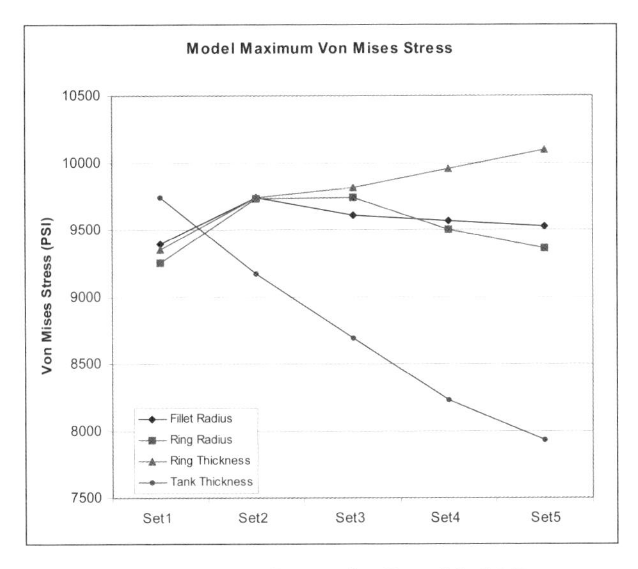


*Figure 10-3: Sensitivity Curves for Four Model Dimensions*

The graphs of Figure 10-3 show that the maximum Von Mises Stress in the model is much more sensitive to the variable Tank Thickness than to the other geometric variables. Consequently, manufacturing tolerances on this dimension should be monitored carefully and additional study might be warranted to ensure that the extremes were within acceptable design limits. Additionally, this graph tells the designer that this is the best place to start if he or she wants to make an impact on the maximum stress level in the system.

If your tool allows you to create these types of studies, they can provide great insight into your designs and should be explored whenever possible.

## 10.4 Correlation to Test

One results interpretation task that is common to nearly all users of analysis is correlation to lab or field data. One of the most effective uses of FEA is to replicate a test that has been proven over time to indicate acceptable production behaviour. Making sure the analysis will give the same answers as the test is an important step in relying on the simulation more heavily. Some general comments on correlating to test data are warranted as it is a high visibility task.

### 10.4.1 How Good is Good Enough?

Inevitably, when analysis results are compared to test results, someone is going to ask the question, "Why doesn't it match perfectly?" Is a 1%, 5%, or 20% difference acceptable when trying to compare test to analysis? There is no easy answer to this question. 20% difference isn't necessarily bad and 1% difference shouldn't necessarily imply accuracy. The best way to understand "good enough" is to consider the scatter in test data itself. In the design of a plastic cover, the engineer scheduled a test specifically to assess the validity of FEA results. When this test set up was critiqued, it was discovered that assembly variance in the fixture and slop added potential variance on the order of magnitude of the measure response. Additionally, when a single part was tested several times with fixture slop accounted for, a 30% variance in the displacement was recorded. Based on all this, if the analysis was within 100% of the mean test deflection, correlation should be assumed. As it turned out, the analytical deflection was within 15% of the mean test deflection. Does that mean the analysis was more accurate than expected? Absolutely not! It simply meant that the model was close enough to draw comparative conclusions from.

While this is one example, it is not an isolated instance. The bottom line is that you shouldn't expect analysis results to correlate any more accurately than reasonable variability in testing. After that, uncertainty in the FE model should help guide you to a reasonable correlation expectation.

### 10.4.2 Get as Close as You Can to Test Set-Up

These other uncertainties in the model, material properties, meshing choices, loads, and restraints, can have a real impact on correlation. The amount of diligence you apply to correlation is directly related to the confidence you can place in your correlation. A colleague of mine was fond of saying, "Test what you analyze, analyze what you test." Examination of test interactions and how they compare to analysis model setup is important.

As the analyst, you should make sure that you observe the test configuration and have the ability to adjust either the test or the analysis to match each other. When given the chance to observe the test and make the necessary adjustments, I've experienced great success in achieving correlation. In cases where I've been asked to explain correlation differences without the benefit of observing the test, it's hit or miss. You can only guess at likely causes for the variance and may never fully understand the differences.

### 10.4.3 What to Measure for FE Correlation

When comparing test to analysis data, the easiest response to measure and correlate is deflection. Since deformation is the primary output of an FEA solution, making sure deformation is correct is also the first step towards correlating stress results. You should measure multiple locations and ensure that fixture deflection or slop doesn't muddy the measurements on the part. Also note any slop in assembly interactions so that these can be factored into measurements. It is worthwhile to attempt to bias an assembly to a position where all slop is taken up before taking measurements you expect to use for correlation since building the slop into the analysis is a more difficult and dubious course of action.

Strain gage measurements are good ways to correlate analysis results. Having the analysis results available to guide strain gage placement is extremely helpful. If you are able to plot P1 vector orientations, these can be used to orient directional strain gages. Try to place strain gages in areas of predictable or gradually changing stress. When strain gages are placed in the anticipated failure zones, typically areas of high stress gradients, it is likely that gage will average the strain across the area of the device. This can result in a large drop in effective strain measured. Additionally, when a gage is placed in areas of high gradient, the measured strain is extremely sensitive to placement so you may not really know what strain is being recorded.

If you can identify areas of pure or near-pure bending, tension or compression, placement of gages in these predictable, placement insensitive locations will tell you if the general model response correlates. This confirms model stiffness & load path. After that, if the geometry in the stress riser is reasonably modelled, the FEA is probably a more accurate indicator of this response and should be relied on to determine part acceptability.

## 10.5 Chapter Summary

You should know going into an analysis what data you'll need to examine based on the goals of the project. Make sure that you understand the data being presented by the software and interpret it in light of all the assumptions made. When a part doesn't break in test after your analysis OK'd it, don't confuse this with correlation. In analysis, two wrongs can make a right and the fact that the part didn't break may have nothing to do with your work up front. Take steps to actually correlate deflections and strains so that you know you are setting up your problems correctly. Only when you have confidence this is correct can you start to replace testing with simulation.

# 11. Introduction to Optimization

The field of optimization warrants a book of its own but design engineers need to know that many optimization tasks are accessible and vital to the process of design.

## 11.1 Basic Optimization Concepts

Optimization is a critical part of the design process. Any design task is an attempt to make something better-safer-cheaper or to provide a better-safer-cheaper solution to an existing problem. To improve is to optimize. A common misconception is that to "optimize" means to engage in a structured investigation with computerized algorithms. The fact is when you tweak a wall thickness, rib placement, or hole size to reduce stress or increase stiffness, you are optimizing. Even if you do use some of the more automated tools within design FEA tools, nothing is forcing you to engage in a lengthy research project. Sometimes, grabbing a couple of dimensions you think might impact the design quality and letting your software optimize on them for a few minutes can provide great insight into why features need to be the way they are. In many optimization tasks, you'll find that initial guesses are pretty good. In others, they might not be. Focus on the insight gained from the task. Achieving the elusive "optimal design" isn't as important as understanding the interactions of features. If you are stuck in the single analysis, "works or doesn't work" mentality, you are missing out on most of the value simulation can provide. It has been said, "It's not the destination but the journey…" This has some meaning in design as well.

I recommend a practical approach to design optimization. While many larger engineering organizations have groups who do nothing but work with obscure and complex optimization tools, often with great success, most design engineers don't have the time or resources so finding a happy medium makes sense. A key aspect of practical design optimization is to change your mindset from the single analysis "Go-No Go" mentality, a static perspective, to a more fluent, design centric mentality. **Instead of asking questions such as "Will it work?" or "Does it break?", use analysis to ask "How thick should my wall be" or "How many bolts do I need?"** The second approach opens your mind up to improvements suggested by the data. The first works for "good enough" but doesn't leave the door open to explore better ways.

## 11.2 Robust Versus Accurate Optimization

One concept to keep in mind when considering practical design optimization is the difference between "Robust" and "Accurate" optimization. These are terms used by optimization specialists but have value in the design world as well. A Robust optimization is one that shoots for the best concept, not necessarily the absolute best configuration of that concept. In the simple example shown in Figure 11-1, a designer working on a square cover may want to know what the best approach for stiffening the plate is. He might consider more fasteners, a rib pattern, or a domed shape. Each of these has potential but which is the best direction to pursue? A Robust optimization task would seek to ascertain the best direction.

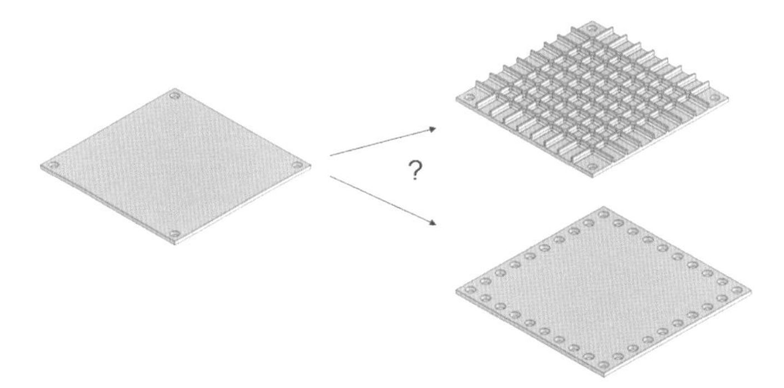


*Figure 11-1: Possible Directions for Stiffening a Plate*

In contrast to this, other optimization approaches might be considered “Accurate”. An Accurate optimization seeks to determine details not direction. Given a plate with ribs, an Accurate optimization approach will return the best rib placement and size (Figure 11-2). However, you’ll get no feedback on whether you should have used ribs in the first place. Consequently, using only this approach, you may come up with the absolute optimal configuration of a bad idea.

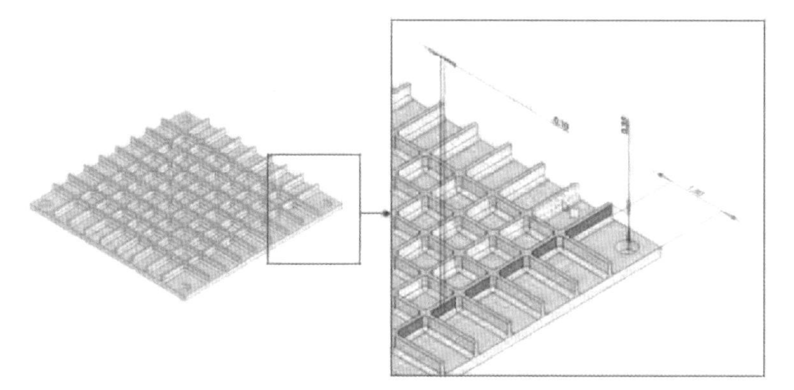

*Figure 11-2: Dimensions for Optimal Sizing of Ribs*

Another way to look at the differences between Robust and Accurate optimization is on the simple one-parameter response curve of Figure 11-3 where variation in a parameter (fillet radius, hole placement, wall thickness) is plotted on the X-axis and the goal of the optimization (weight, stress, displacement) is plotted on the Y-axis.

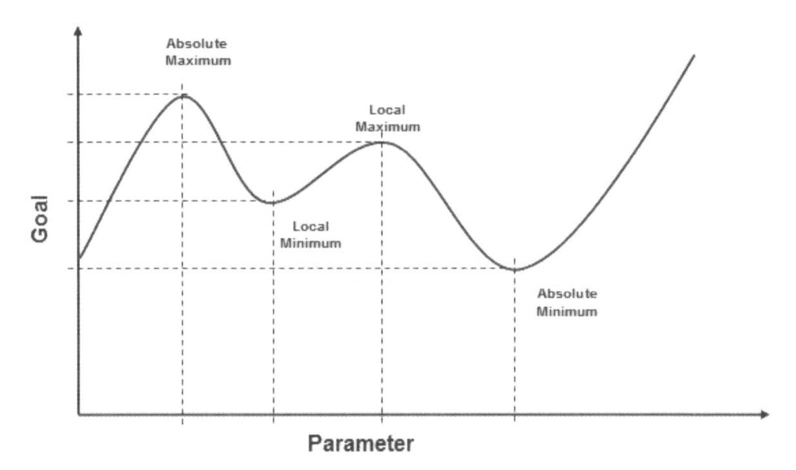


*Figure 11-3: Single Parameter Response Curve*

As can be seen from the curve, there are multiple peaks and valleys in the parameter space. The bottom of each valley is called a minimum and is clearly a better parameter value than the neighboring values. However, in this response curve, there is clearly an absolute minimum, a "best" parameter value.

Robust optimization schemes look at the big picture to find regions of absolute max & min response. A robust algorithm or process may not return the precise value that is the absolute minimum (in this case) but will get you in the right "valley".

Accurate optimizers are typically the realm of gradient based routines that evaluate the impact of small changes to an initial value. When the optimizer finds that the response worsens when the parameter value is increased or decreased, based on the defined goal, it will typically return that design point as optimal. These type of algorithms are considered "starting point sensitive" in that they might come to different conclusions based on the initial configuration of the model.

Most analysis tools provide ways to perform both robust and accurate optimization tasks on your designs, although you won't see them labeled as such. The rest of the chapter will explore techniques such as Optimization & Trade Studies to point you in the right direction (robust) and Sensitivity Studies to fine-tune the configuration of a concept (accurate). Additionally, the importance of proper model preparation to keep all your options open will be discussed.

## 11.3 Model Preparation for Optimization

One of the biggest hindrances to more pervasive optimization (or design analysis for that matter) is time. If it takes too long or is too cumbersome, most design engineers will opt for the "gut feel" design and deal with the consequences if it isn't "good enough". Putting some forethought into the set up of your CAD model and preparing it for design analysis and optimization can really help reduce the time and effort required to run, modify and re-run a study. In the world of single-analysis Go-No Go design, today's solvers are fast enough that it may be quicker to run a poorly prepared CAD model through a static study than spend time reducing it. However, the nature of optimization is to run multiple iterations and explore alternatives. From that perspective, model size reduction becomes important.

### 11.3.1 Reduce Model Size

One of the steps you can take to reduce model size is to use symmetry, where applicable, to get the same results on a portion of your model as you would have gotten on the whole thing. The example shown in Figure 11-4 has 3 planes of symmetry allowing only 1/8 of the model to be solved without any loss of accuracy, provided the boundary conditions have the same symmetry.

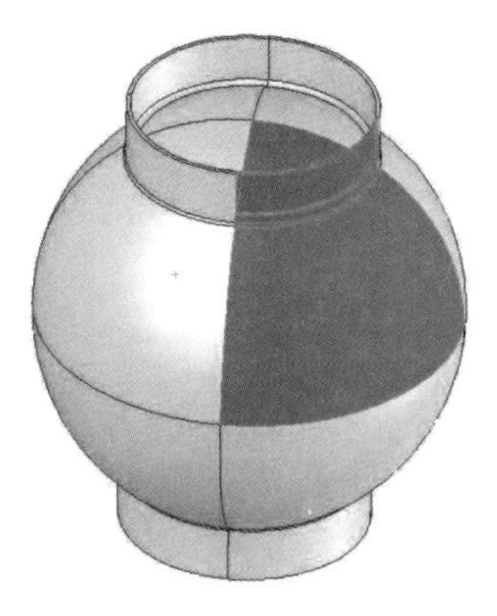

*Figure 11-4: Full Model of Pressure Vessel Indicating Symmetry*

Feature reduction & simplification is another important approach. As discussed in Chapter 7 "abstinence" in design is advocated over "suppression". Don't even put a feature in the model you would expect to suppress later. Verify the macro level behavior using only the important features and add in others as the analysis suggests or for final design documentation. Consider alternate idealizations such as shells and beams where applicable and virtual structure options in your software to reduce overall model size.

### 11.3.2 Feature Creation

Another important model preparation task is to explicitly dimension any feature that might benefit from optimization. This is a restriction of the technology within most feature based optimization tools. The automated routines that vary a feature size typically have to work with dimensions. If you didn't define them in your feature creation, you won't be able to control them. Also consider using more general geometry construction methods that minimize dependencies and feature complexity. Add fillets explicitly vs. building them into the sketch. Try to dimension using datums vs. other features so you don't accidentally change a child while optimizing the parent. Put fewer entities in feature selection sets. Grab functional groupings to fillet in one feature that you would always expect to stay equal. This again minimizes the chances you'll change something unintentionally when optimizing.

It's safer to fully define sketches and assembly mates. While most CAD systems provide tools to free-form your concept, there is nothing more frustrating than to accidentally move a part due to a screen pick when trying to tweak a design. You can easily draw an invalid conclusion if the results change because of the accidental move than the intentional variation. Finally, explore the extents of any geometric changes you are considering in CAD before trying to control them with automated optimization tools. If these automated routines encounter a conflict toward the end of a long run, you may have to start over.

### 11.3.3 Meshing

You may not be aware of it but mesh variations from iteration to iteration can cause as much variance in the results as small geometry changes you might be exploring. The example shown in Figure 11-5 is a turbine for a dental hand piece which spins at high speeds. The designer wanted to see what stress might be generated if the worst imbalance possible based on manufacturing tolerances was introduced. The initial studies yielded an inconsistent scatter of data and it wasn't until the mesh was forced to be nearly identical from iteration to iteration that the true nature of the variation was understood.

*Figure 11-5: Dental Hand Piece Turbine under Centrifugal Loads*

The safest and easiest approach to achieving consistent results is to use aggressive convergence practices and automatic adaptive meshing, if available, for all studies. Most designers don't make changes on such a small scale that minor mesh variations will have the impact as in the previously cited example but failure to converge or inconsistent convergence can result in large stress deviations.

### 11.3.4 Baseline Design Validation and Improvement

The first step in any optimization process is to determine if your design satisfies operational specs. While optimization techniques can be used to identify direction, they should not be used to re-design a concept that doesn't work. The most efficient use of optimization is to improve a working design, either in weight, cost, stress, or stiffness. If your first pass design doesn't satisfy all the applicable allowables, use Trade Studies, or "What If" studies to explore alternate concepts. Try many different concepts and approaches and focus on insight, not necessarily accuracy. At this stage, you aren't looking for small percentage point improvements since it is unlikely that these will be meaningful compared to all the other uncertainty in the model (loads, material properties, geometric variability.) Instead, look for improvements on the 50% or better level. These are real differences that will stick as you work with the design. Consequently, worrying about model precision isn't as important as modeling consistency.

Finally, explore big changes in pursuit of big improvements. Small dimensional tweaks are not as productive as a big modification. If you are considering improving stress by increasing a fillet radius, make the radius as large as you can justify in the first pass. If this doesn't fix the problem, a somewhat smaller change won't have worked either. Use these big changes to identify features and dimensions that have the biggest impact on the results of interest, the changes your part are most sensitive to.

Another benefit of this technique is that you can quickly rule out certain "obvious" fixes that you or others might suggest instead of spending time and risking emotional commitment to the wrong approach.

Trade Studies provide a Robust approach to optimization or design improvement. Sensitivity Studies, as introduced in the previous chapter, provide an "accurate" follow on to the directions learned in the Trade Studies.

In the simplified pressure vessel example shown previously in Figure 11-4, a Sensitivity Study was used to identify the optimal wall thickness that brought tangential stresses right to the allowable stress. Some of the results of this study are shown in Figure 11-6.

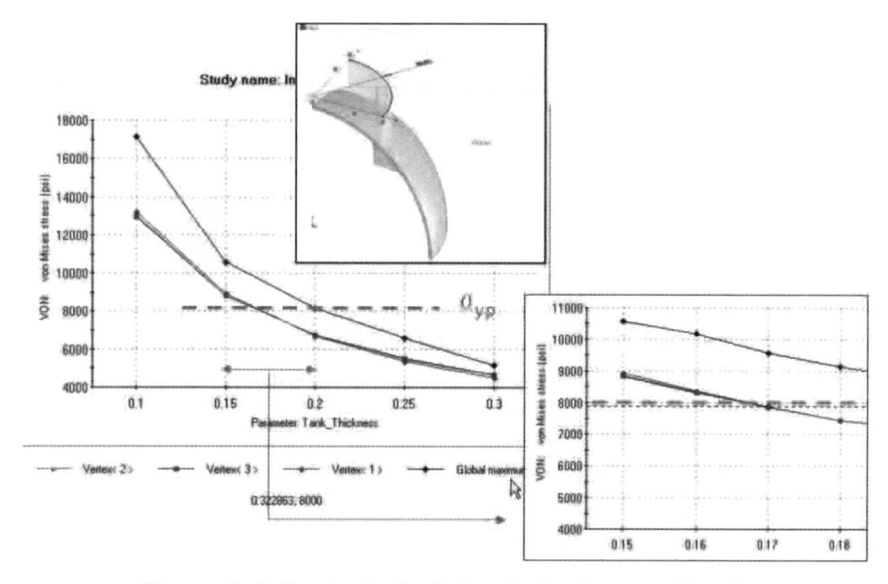


*Figure 11-6: Sensitivity Study Results for Pressure Vessel*

Big steps were used in an initial pass where the thickness was varied from the minimum thickness that could be manufactured to the maximum thickness the application could tolerate. It was seen that the curve crossed the allowable stress threshold somewhere between 0.15 and 0.20 inches. A second, more refined, study explored the response in that smaller range to suggest that a wall thickness of 0.17 inches was optimal.

A third tool for optimization in FEA, after Trade Studies and Sensitivity Studies, is an actual Optimization Study. All automated optimization routines have the same 3 basic building blocks: The Objective, Constraints, and Design Variables. In short, the optimized design should maximize or minimize the objective by changing the variables while keeping certain responses within the defined constraints.

The solution time and resource usage of an optimization study is directly dependent upon the number of Design Variables. It is in your best interest to minimize the number of variables provided to the optimizer by taking advantage of the lessons learned from Trade or Sensitivity Studies.

The design variables you choose to provide to the optimizer must be "smooth variables" which means the range of possible values must include all possibilities within the upper & lower limits. The other type of variables you might want to consider are "discrete variables" which have incremental or pre-defined values within an upper or lower limit and values in between these increments are not possible or realistic.

For example, the height or width of a rib in a plastic part can be smooth variables. Contrast this with the number of ribs in a feature. If you can fit 1 to 5 ribs in an area of your part to stiffen it, your choices are 1, 2, 3, 4, or 5 but not 2.23432. Another type of discrete variable is the "On/Off" condition where a feature such as a rib, a weld, a stiffener, or a fastener is there or it isn't. Typically, discrete variables can't be used in an automatic Optimization Study but you can explore these with Trade Studies and, to a lesser extent, Sensitivity Studies.

There is another class of variables that called "Semi-smooth" that can be used in automated optimizers. These include inputs such as sheet metal gage or bolt sizes. While these have fixed increments, the possibilities are comprehensive enough that it is reasonable to treat them as smooth and then round the optimized result up or down.

Once all three building blocks are defined, the optimizer iterates with the available data. If a design that satisfies the constraints is found, the optimization will have succeeded. If the optimization fails, either the dimensional variables conflicted and the software couldn't construct a valid model or a condition satisfying the constraints couldn't be found.

When the optimization completes successfully, you'll be provided numerous plots that show the history of all the iterations explored in the study. These plots can suggest which variables the optimizer spent the most resources on, indicating again which might provide the biggest impact on future improvements. The most telling result is the final configuration determined by the optimizer. Depending on the integration with your CAD system, this optimized design may be automatically reflected in your parametric solid model.

## 11.4 Chapter Summary

It is reasonable to expect a 15% to 20% improvement in weight from a moderately aggressive optimization program on a system that hasn't been optimized, either using software or manually, in the past. The insight gained and the lessons learned about your design, simply by pursuing optimization, is as valuable as the optimized design itself. This insight can be applied to future designs so that the initial concept is closer to optimal and has truly benefited from prior experience.

# 12. Project Reporting

In the course of a design project, it is conceivable to analyze hundreds of parts and configurations. Tracking all the changes that were made and the data used to suggest these changes can be a daunting task. However, most design FEA tools provide varying levels of automatic or semi-automatic report generation that can help you document the work you did. Despite the added step and the perceived hassle, documentation is important for a number of reasons and you should find a process or format that works for you to document sufficiently without it being too great a burden.

## 12.1 Importance of Reports as a QA tool

If nothing else, writing a report may be the single best quality assurance (QA) tool available to an analysis user at any skill level. I can recall countless times where the exercise of documenting my steps and explaining my conclusions indicated holes in my logic or mistakes in units or assumptions that required me to clean up loose ends before issuing formal conclusions based on my results. Actually, automated tools take a little of this QA benefit away because the analyst isn't required to compile and format the data manually. However, proofing the report which shows the data in a different, more concise format, still allows the analyst to review the program in its entirety from a more detached viewpoint which should help to highlight inconsistencies in the model or the project flow. It is usually helpful to have a colleague review your report for completeness, both in the project and in the reporting of the project, before issuing it.

## 12.2 Documentation for Posterity

The most obvious reason to complete a report on analysis tasks is to provide documentation of what was done, why it was done, and what the results & conclusions were for anyone who might need to investigate the project later. In many cases, a design engineer or analyst may need to revisit an analysis done previously, either because a question on the data has arisen such as a field failure, or because a similar system needs to be analyzed and it makes sense to use the same setup for consistency. In other cases, projects or products get passed to other designers or engineers who have to decide if the conclusions drawn from previous analyses, or even those analyses themselves, were valid. I have observed cases at companies of all sizes and in nearly all industries where absence of documentation from previously completed projects has forced duplicate effort because the new design team couldn't be sure that the prior project was analyzed with the best techniques. A good report could have answered all the questions these verification studies addressed.

## 12.3 Minimum Content of a Project Report

Many design engineers consider screen shots pasted in a text document a "report". When reviewing these "reports", even the engineers who created them get frustrated trying to remember which results went with which iterations. At a bare minimum, a report should include these sections:

- Goal of the study
- All assumptions used in completing the study
- Geometry used including any simplifications
- Material properties and, if possible, the source of the properties
- Loads applied and the source of or justification for those loads
- Restraints used and some justification of those restraints
- All results of interest clearly labelled with the geometry iteration they refer to
- A summary of the meaning of the results as they relate to the goal
- A conclusion that directly addresses the goal

When writing a report, more is better. Try to be concise but don't neglect important data. In the end, someone who is familiar with the product analyzed should be able to gauge the validity of the assumptions used without having to repeat the analysis.

## 12.4 Quick and Dirty Reporting

This seems like a lot of work in an already overloaded project schedule. However, there are ways to simplify and speed up the process. First of all, take advantage of any automatic report generation tools your software provides. If there is any customization available to minimize the editing you may need to do on individual reports, make sure you spend some time getting that the way you want it. Next, output the report to the most accessible format your software allows. Most tools offer an HTML report generation format. While I don't know where that trend started, I can say that this format may be one of the least user-friendly formats if any data needs to be added or the report needs to be modified. Many tools now offer a Microsoft Word option for reports. This seems to be more widely used since the modification options are nearly limitless.

Once you have your base report, it should only take a few minutes to type in notes on material or load sourcing, goals, and conclusions, if the report builder in your software didn't give you placeholders for these. While better formatting makes for easier reading later, in a pinch, I'd suggest getting the data in as completely as you can and if formatting has to suffer because of time restrictions, at least the information is documented.

Another report format that I've found helpful when in a hurry utilizes Microsoft PowerPoint, although any free-form tool that allows sequential pages would do. I'll often keep PowerPoint open in the background. As I'm building a model or reviewing results, I'll use a screen capture tool to copy and paste images of geometry, material definition forms, load & restraint definitions, run summaries, etc... I'll type in notes over an image, draw arrows, circles and boxes to emphasize an area as shown in the example of Figure 12-1.

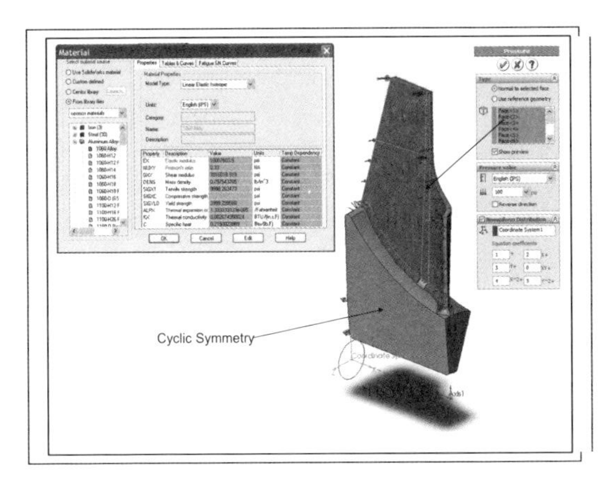


*Figure 12-1: Example of Quick Report Format*

## 12.5 Chapter Summary

While it may seem like there is never time to write reports, finding a format that works for you, even if it is quick and dirty, will end up saving you time in the long run. The benefits from both documentation and quality checking of a project are real. As you get more comfortable documenting your work, this will become clear and the process will become second nature.

# 13. Where to Go For More Help…

To conclude this reference for design analysts, it's fitting to offer some suggestions for on-going education and support. No single text can provide all the information you need to use a technology as deep as FEA to its fullest extent and that was not the purpose of this book. While many design analysis users never look for more education beyond their intro to software class, the fact that you've made it this far suggests that you are open to additional sources of information. The following is a collection of resources that you may find useful as you grow in analysis.

## 13.1 Brief Discussion of the FE World Outside of CAD Software

While the stated focus of this book was on CAD embedded FEA tools, the underlying technology behind all of the most popular tools is rooted in a full-featured analysis tool that most likely still exists, is supported and used by analysis specialists or design engineers who pride themselves on doing things the hard way. Many short-cuts in a design analysis tool actually create a combination of analysis entities in the native format of the root software.

The more you understand the root code, the basic building blocks that have been simplified and automated with the CAD-embedded UI, the easier it will be to understand why some features behave as expected and why others don't quite seem to represent the real structure they imply. Additionally, you will be surprised at how many options and capabilities exist behind the GUI and knowing of them may suggest different ways to approach current problems.

Your software supplier's tech support should be able to point you to documentation that helps you understand the underlying FEA technology. In fact, they'll probably welcome the effort since it will make you a more informed user. Additionally, there are a number of texts on general FEA, listed later in this chapter, that provide information that is applicable to nearly all analysis codes which can provide great insight into why the software behaves the way it does.

## 13.2 Overview of NAFEMS Documentation

In addition to this document, NAFEMS offers the most complete library of software independent FEA-related literature. Three of my most recommended books are:

- NAFEMS - The Finite Element Primer
- NAFEMS – Introduction to Nonlinear Finite Element Analysis (Hinton)

- NAFEMS – A Finite Element Dynamics Primer (Hitchings)

A review of the NAFEMS website (www.nafems.org) will yield numerous other texts, including "An Introduction to the Use of Material Models in FE." which is referred to above, and publications that may be applicable to your product development needs. The NAFEMS magazine, BENCHmark, is published quarterly and contains articles on all types of analysis with an emphasis on case studies and practical applications. The NAFEMS Benchmarks, the namesake of the magazine, are a series of test problems covering all aspects of FEA including linear static, nonlinear, dynamic, fluid flow, and contact modeling to name a few. These are excellent learning tools since, for each simple problem, a solution is provided to check your work. They allow you to focus on the basic analytical techniques without being burdened with complex or lengthy models.

## 13.3 Internet Resources

Of course, it goes without saying, that the NAFEMS page (www.nafems.org) should be included in any list of important FEA related sites. However, there are several other excellent sites worth mentioning. This section has to be prefaced with a warning that anything on the Internet is temporary in nature. All the sites listed below were functional at the time of this writing and have provided support for scores of analysts and designers over the years.

**www.fatiguecalculator.com** – An extensive repository of fatigue information

**www.caddigest.com/subjects/cae/fea.htm** - A compilation of FEA related articles from numerous publications and sources

**http://ocw.mit.edu/OcwWeb/Global/all-courses.htm#MechanicalEngineering** – Open courseware from MIT (Massachusetts Institute of Technology) with course notes on a variety of FEA and mechanical engineering related topics

**www.dermotmonaghan.com** – A comprehensive FEA-related website with articles and links covering nearly all aspects of FEA

**www.polymerfem.com** – A comprehensive website discussing all aspects of plastic and polymer related FEA

www.matweb.com – A comprehensive on-line material database with links to many FEA tools

Between these and the links contained with them, interested users should have no problem locating information on nearly any FEA related subject.

As stated at the beginning of Chapter 10, it is equally important to understand the engineering principals and failure mechanics behind your products and chosen materials. For these areas of knowledge, regular searches on relevant keywords can often produce unexpected and valuable finds that can change the way you think about design. I recommend that everyone make time to investigate this resource.

It is also important to note that NAFEMS is not affiliated with any of these websites and the information provided should be used at your own risk since the accuracy of the information can neither be validated nor discounted to any reasonable standard.

## 13.4 Other Reference Sources

This last list represents references the author uses and recommends regularly. They are always within arm's reach when engaged in design analysis.

- Building Better Products with Finite Element Analysis – Adams & Askenazi; OnWord Press; 1999 (For obvious reasons…)
- Mechanical Engineering Design – Shigley & Mishcke; McGraw-Hill
- Roark's Formulas of Stress & Strain – Young; McGraw-Hill
- Structural Analysis of Thermoplastic Components – Trantina & Nimmer; McGraw-Hill; 1994
- Machinery's Handbook – Oberg et al; Industrial Press Inc.
- Collins, J. "Failure of Materials in Mechanical Design; Analysis Prediction, Prevention" John Wiley & Sons, New York, NY, 1993
- Weaver, M.A. 1999 Determination of Weld Loads and Throat Requirements Using Finite Element Analysis with Shell Element Models - A comparison with Classical Analysis. *Welding Journal Research Supplement* Vol. 78, No. 4, p. 116s to 226s
- Dong P, et al. "A Mesh-Insensitive Structural Stress Procedure for Fatigue Evaluation of Welded Structures" International Institute of Welding, IIW Doc. XIII-1902-01/XV-1089-01, July, 2000
- Lotsberg, I "Fatigue Design of Plated Structures Using Finite Element Analysis" Ship and Offshore Structures 2006 Volume 1, Issue 1 Pages 45-54

## 13.5 Importance of Mentoring or Access to Expert

As a final comment on additional resources, it must be stressed that even the most diligent research can never replace the benefits of working with someone who has been through the learning process and is working successfully with higher level and more complex analyses. If you are fortunate to have a resource of this nature within your organization, do your best to develop a mentor-like relationship with them. In companies that actively pursue mentoring between specialists and design engineers who use FEA part-time, the overall quality of FEA use improves exponentially. Not only do the design engineers benefit from the experience of the specialists but nothing cements and refines knowledge like teaching. I've found that engineers who take the time to support or teach more recent practitioners are some of the most knowledgeable and grounded analysis users. I can also state that they didn't necessarily start out that way but became so as a direct result of their mentor position.

If you don't have access to an internal mentor, investigate external resources to provide this role. The NAFEMS website lists member consultancies so this is a good place to start. Remember that not everyone who uses FEA as a consultant, even those with many years of experience, is as knowledgeable as they think they are and even many of those who are truly experts in their fields are not qualified or cut out to teach. Be prepared to ask for references and don't hesitate to question all of their methods or dictums. If they can't explain their rationale clearly, it's possible they don't really know why they do what they do. That might indicate its time to move on. When you do find a resource who can help you grow, utilize them to review your work and suggest alternate approaches.

## 13.6 Summary of Document

To conclude this discussion of design analysis, the author would like to point out that success with this technology, as with any other engineering discipline, is not guaranteed. While some might project a future where 3D models of entire products can be put through virtual testing with the same ease and robustness as lab testing but with the cost and speed benefits of computer simulation, those days are far off. Software has made the technology more accessible and with that comes the responsibility to understand what you are actually doing. Following the practices defined in this text will provide a good platform for the design analysis most readers will need to perform but all the possible variations and needs that might arise could never be predicted. Question all assumptions and results. Discipline yourself to dig when the answers aren't obvious and explore alternate approaches and you'll find that the field of design analysis will open up many avenues and design options you never before considered.